EXPLORING LASER LIGHT

Laboratory Exercises and
Lecture Demonstrations
Performed With Low-Power
Helium-Neon Gas Lasers

EXPLORING LASER LIGHT

by T. Kallard

Reprinted by American Association of Physics Teachers

DEDICATED TO
THE MEMORY OF MY PARENTS

EXPLORING LASER LIGHT

Copyright © 1977 by Optosonic Press.

Cover illustration by Jean Larcher,
(''Geometrical Designs & Optical Art,''
Dover Pictorial Archive Series, 1974).

Reprinted by:

American Association of Physics Teachers
Publications Department
5112 Berwyn Road
College Park, MD 20740, U.S.A.

Fourth Printing: October 1989
Third Printing: July 1985
Second Printing: January 1982

ISBN # 0-87739-004-5

Library of Congress catalog card number: 70-160227

PREFACE

The purposes of this book are, first, to describe some of the most recent experiments in optics, and second, to point out possibly fresh and simplified ways of performing some of the older ones. It is by no means a comprehensive manual of all the known optics experiments but rather a selective one. It does, nevertheless, explore a wide and far ranging variety of optical bypaths, and thus offers the reader the choice of following those which interest him most.

For example: the emergence of lasers. Many optics experiments used to be difficult to perform with ordinary light sources. With lasers this is no longer so. And when lasers were combined with large-screen closed circuit television, many old and difficult optics demonstrations became easier to do and able to be viewed by a large audience. The low-power, continuous wave helium-neon gas laser is now one of the most dependable tools of the serious experimenter in optics.

The He-Ne laser has also earned a place as an aid to the science educator. It can be used at all levels in education - in the basic high-school physics laboratory as well as in the most complex of university installations. This book explores the properties of the He-Ne lasers and applies them to experiments in optics for use in lecture demonstrations, teaching laboratories and home study experimentation. It will suggest and point the way rather than offering the reader a cookbook type manual. Relevant formulas are presented to refresh the memory and, where appropriate, numerical values for parameters are given. Schematic drawings are used profusely to further help the experimenter.

The problem of cost has not been overlooked. The experiments to be found here were designed to keep the cost of required apparatus at a minimum. Many components are available as surplus items and the experiments were selected on a qualitative basis, this in order to expose the reader to the largest number of optical principles and techniques in the space available. It is hoped this presentation will whet the reader's appetite and tempt him to delve more deeply into textbooks and journals for further study.

The "References" and the literature suggested "For Further Reading" make it easier for the reader to pursue any of the topics covered here in greater detail. Review and tutorial articles are often included in the references. Relevant books are listed at the end of this volume in the extensive Bibliography. Author, title, publisher and year of publication are given.

If this book proves useful, similar volumes will be compiled as new experiments are devised and new developments emerge. Comments and suggestions from readers are welcome and should be addressed to the publisher. Contributors of new experiments will receive full credit if their work is incorporated in later editions.

The author thanks the many people who gave advice, help and information in the preparation of this book. Especially does he wish to thank the manufacturers who so generously furnished equipment, raw materials, data and illustrations.

T. KALLARD

CONTENTS

"""""""""

INTRODUCTION

What is a laser?

The term "laser" is an acronym for "Light Amplification by Stimulated Emission of Radiation." Thus, the laser is a device which produces and amplifies light. The mechanism which accomplishes the stimulated emission was postulated by Einstein in 1917. Lasers may generate energy in the ultraviolet, visible, or infrared spectrum. The first continuously operating (c.w. - continuous wave) helium-neon laser was reported in February 1961 by Javan, Bennett and Herriott of the Bell Telephone Laboratories. Helium-neon lasers produce an intense, coherent, visible light beam of wavelength 6328 A (Angstroms), or, expressed in another unit of length: 632.8 nm (nanometers). All exercises and lecture demonstrations contained in this book utilize low-power c.w. He-Ne (helium-neon) lasers.

How does the He-Ne gas laser operate?

Without delving into the mathematics and quantum theory involved in the operation of a laser, the simplest way to describe the device is to compare it with an electronic r.f. oscillator.

An electronic oscillator (Fig. 1) has four main parts: (1) amplifier, (2) resonant feedback network, (3) output coupling port, (4) power source. The corresponding parts of a laser are shown in Fig. 2. Here the amplifier is a glass tube which contains a gaseous mixture of helium and neon, with neon as the active lasing material. When the laser's power supply (the "pump") delivers enough energy to cause continuous glow discharge in the gas tube (much the same as a neon sign is pumped by an electrical discharge), the neon atoms are elevated to a higher energy state by colliding with the helium atoms. When the neon atoms drop back to their lower energy state, they give up energy at certain wavelength: in this example the wavelength is 632.8 nm, in the red portion of the visible spectrum. The light output will be random and scattered equally in all directions. Some of this light is

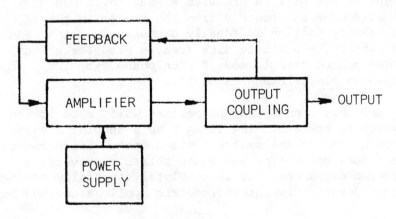

Fig. 1

lost through the side walls of the glass tube, but the portion that travels down the center of the tube strikes other excited neon atoms creating more light energy of the same wavelength.

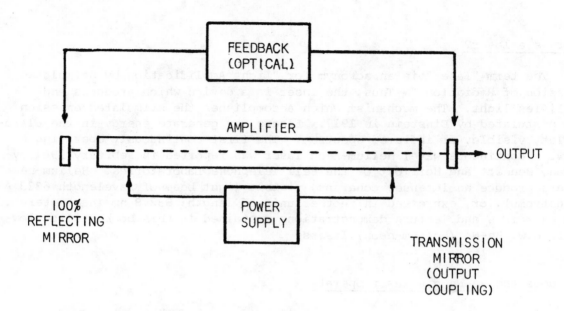

Fig. 2

The laser tube is placed in an optical cavity formed by two highly reflective mirrors positioned to face each other along a central axis. The mirrors reflect the initial beam which, as it bounces back and forth, eventually builds up enough energy to emerge through whichever mirror has the least reflectivity. This escaping light constitutes the highly directional beam of the laser. Since in the He-Ne laser, light amplification is only 1.02 on each pass of the beam from one mirror to the other, all losses must be kept below 2%. The so-called transmission mirror is coated to allow less than 1% of the generated light to escape. Thus, the beam emitted is less than 1/100th as intense as the beam between the mirrors.

Since the laser tube and the mirrors form an optical resonant cavity, the optical path length between successive reflections at a mirror must be of an integral number of wavelengths to produce reinforcement of the wave. Perfect alignment of the mirrors produces a beam which has an irradiance distribution that decreases smoothly from the center to the edge of the beam; the flux density pattern is ideally Gaussian over the beam's cross-section. This pattern, - a single disk area, - produces a single spot of light. It is designated "single mode," "uniphase," or TEM_{OO} mode. The last from Transverse Electric and Magnetic.

The TEM_{OO} mode has a number of properties which make it the most desirable mode in which to operate. The TEM_{OO} beam's angular divergence is smallest and can be focused down to the smallest sized spot. Furthermore, the TEM_{OO} (uniphase!) mode does not suffer any phase shifts or reversals across the beam as do higher order modes. It is completely spatially coherent. This is an important consideration in interferometric applications and holography.

Properties of laser light.

Laser light is quite different from light normally encountered. It has four unique characteristics that make the device a useful tool: (1) it is highly directional, (2) coherent, (3) very bright, and (4) monochromatic.

1) The directionality of the laser light is because only the light on the axis between the mirrors can escape from the laser. The beam emerges inherently well collimated and highly directional, and thus useful for applications where high concentration of light in a given direction is important.

2) The coherence of laser light in time and space is the one previously unobtainable property that makes it such an important source of light. Only light whose multiples of half a wavelength fit exactly between the mirrors is allowed to escape from the laser. Thus, standing waves are established between the mirrors, and each light particle is in step with the others. Since the light produced by the laser can be thought of as a wave oscillating some 10^{14} times a second, for such a wave to be coherent two conditions must be fulfilled: 1) It must be very nearly a single frequency (the spread in frequency or linewidth must be small). If this condition is fulfilled, the light is said to have temporal coherence, and 2) the wavefront must have a shape which remains constant in time. (A wavefront is defined to be the surface formed by points of equal phase. A point source of light emits a spherical wavefront. A perfectly collimated beam of light has a plane wavefront.) If this second condition holds, the light is said to be spatially coherent.

3) Intensity and monochromaticity go hand in hand. Since the laser builds up energy of only one frequency, all its power per interval of wavelength is much greater than the power available from other sources, this is simply because of its greater monochromaticity.

4) Monochromaticity (single coloredness) is the result of the narrow pass band of the amplifier plus the selectivity of the resonant feedback mirrors. For example, the wavelength of the red light emitted by a He-Ne laser is 632.8 nm. It is possible to limit the wavelength spread to a small band, say from 1 nm to 10 nm and produce light of high chromatic purity. Such light is called roughly "monochromatic" light, meaning light of a single color. If we refer to monochromatic light of 633 nm, it means a small band of wavelengths around 633 nm.

The anatomy of a He-Ne gas laser.

A) The Plasma Tube.

The plasma tube is a long capillary tube, two millimeters in diameter, surrounded by a hermetically sealed outer tube, one inch in diameter. The laser action which produces the beam occurs in the central capillary tube as the high voltage D.C. is applied to a mixture of gases, approximately 85% helium and 15% neon, at a pressure of about 1/300 of an atmosphere. As the electric energy is applied, the electrons of each atom respond by changing their orbits from the normal ground level configuration to the larger and more complex orbits associated with higher energy levels. After a short time in the energized state, the electrons spontaneously revert to their original conditions and the characteristic spectra of both helium and neon gases may be observed as each of the atoms radiates its recently acquired energy. This produces the characteristic blue light of helium gas and the familiar red glow of neon which may be observed in the laser tube. IMPORTANT NOTE: <u>Do not look into the laser!</u> Hold a piece of thin paper in the beam path near the exit window of the laser to observe the colors.

Although neon radiates several different wavelengths of light as its electrons fall from higher energy levels to the ground state, one of the strongest radiations in the visible light range (632.8 nm) is produced when the orbital electrons fall from the $3S_2$ to the $2P_4$ level. When one of the neon atoms undergoes this particular transition, a photon of light travels down the laser tube and other energized neon electrons along its path are stimulated to undergo the same transition. This frequency action produces additional radiations of the same frequency. The phenomenon is called stimulated emission or radiation. The stimulated emission results in a combined wave of increasing amplitude. Upon reaching the end of the laser tube, the wave encounters a mirror which sends it back through the tube to stimulate more energized neon atoms and increase its amplitude by a factor of 1.02 with each pass. With a flar mirror at each end of the laser tube, perfectly aligned waves of high amplitude are generated in a very short time.

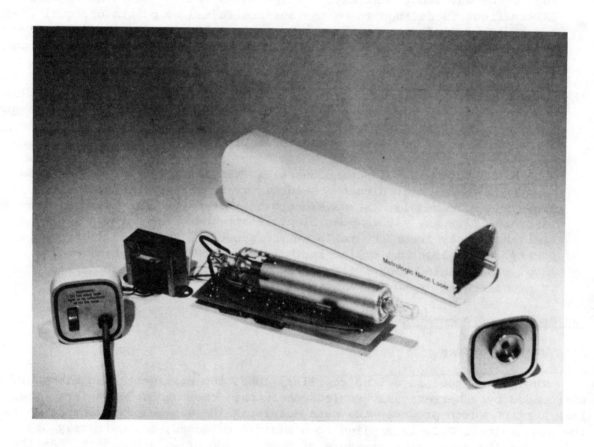

Fig. 3. Exploded view of laser

These waves are coherent in time because only those waves with an integral number of half wavelengths from mirror to mirror can sustain oscillation.

The situation is similar to the standing waves in a jump rope.

To produce an external laser beam, the mirror at the front of the laser tube is a partial reflector which reflects 99% of the light and transmits approximately 1%. The mirror at the back end of the laser tube has a higher reflectivity and reflects about 99.9% of the light while transmitting less than 0.1%.

During the manufacturing process, the coatings of the two mirrors are carefully adjusted so that the laser will resonate at 632.8 nm emission at the expense of other radiations produced by the neon gas.

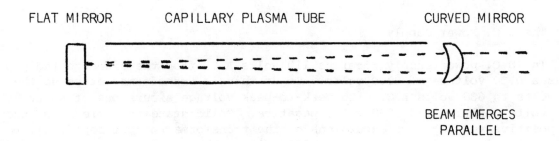

FLAT MIRROR   CAPILLARY PLASMA TUBE   CURVED MIRROR

BEAM EMERGES
PARALLEL

Fig. 4. The capillary plasma tube

A "semi-confocal" mirror arrangement is used in the plasma tube (Fig. 4). This consists of a flat mirror at the back of the laser tube and a concave mirror at the end where the beam emerges. Although a greater power output could be obtained with a flat mirror (or a long radius curved mirror) at each end of the laser tube, flat mirrors are very difficult to align; it is even more difficult to maintain their alignment when the laser is subjected to minor mechanical stresses during operation. With the semi-confocal arrangement, some power is sacrificed but the laser is so stable it can withstand the rough vibration and stress which occurs in a typical student laboratory. Furthermore, the curvature of the mirror at the output end of the laser tube is calculated so it will focus the laser beam at approximately the plane of the distant flat mirror. This curve/flat arrangement produces a laser beam which is cone shaped between the mirrors, the point being at the flat end, and diverging at the curved end. To compensate for this divergence, an additional converging lens surface is placed on the laser output mirror to produce a beam whose edges are very close to parallel.

Because of the internal geometry of individual laser tubes, it is found that the beam tends to vibrate more strongly in a particular plane than at any other possibilities. That is, the beam tends to be elliptically polarized. It is also observed that there is sometimes a secondary beam, polarized at right angles to the favored direction of vibration. In a short laser tube, one will find that the output beam is polarized at a given instant and that this plane of polarization appears to shift between two favored directions at right angles to each other in a somewhat unpredicatble manner. This interesting effect may be observed by passing the laser beam through a polarizing filter and observing the changes in beam intensity.

The capillary tube in which the laser action occurs is surrounded by a second tube about one inch in diameter. This outer tube has two purposes:

1) It supports the inner capillary and the two end mirrors in a rigid permanent alignment, 2) it provides a large   reservoir for the neon gas which replenishes the supply in the laser cavity as it is slowly absorbed by the cathode during laser operation.

Helium gas is included in the laser because it has been found to enhance the output of the neon gas by a factor as high as 200x. As the helium atoms are energized by the high voltage direct current, they collide with nearby neon atoms in a most efficient energy transfer process. Although it has been found that the neon gas alone will provide lasing action, the output is about 200 times as great when helium and neon are mixed in proportions of about 6 to 1 (i.e., about 85% helium and 15% neon).

B)   The D.C. Power Supply

The D.C. power supply receives 110 volts A.C. from the linecord and produces a D.C. voltage of 2000 volts. To do this, a transformer steps up the 110 volts to 630 volts A.C. with peak-to-peak voltage excursions of about 1000 volts positive and 1000 volts negative. Solid state rectifiers act upon the positive and negative excursions of the transformer output separately to produce two independent outputs of 1000 volts. These voltages are then added in series using a voltage doubler circuit to produce a combined output of approximately 2000 volts. This is reduced to the required 1100 volts with the aid of a string of ballast resistors. To start the initial laser action and ionize the gas in the tube, a separate circuit provides a pulse of about 2000 volts which is automatically removed when the laser action starts.

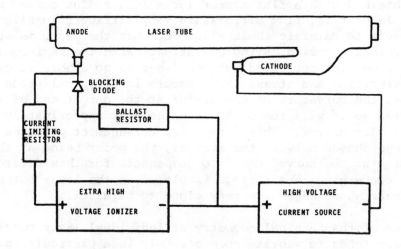

Fig. 5. Gas laser power supply

References:

Knowles, C.H., Popular Electronics p27 (Jan '69)
Gottlieb, H.H., "Experiments Using a He-Ne Laser," Metrologic Instruments,
          Inc., Bellmawr, N.J. (1971)
Kruger, J.S., Electro-optical Systems Design p12 (Sep '72)
H. Weichel and L.S. Pedrotti, Electro-opt Syst Design p22 (Jly '76)
H. Weichel, Am J Phys v44 p839 (Sep '76)

Powerful pulsed lasers are used for welding and, also, are capable of drilling holes in metal. High-power lasers are used in surgery, in resistor trimming, in non-destructive testing, and even in static and dynamic art exhibits. More and more graduate research is performed with high-power lasers in the 10^2 - 10^{12} W/cm^2 range. Therefore, it is essential that all persons who are exposed to laser hazards be informed on the subject of laser safety.

All exercises and lecture demonstrations described in this book can be performed with lasers whose output falls in the 1-5 mW range. A recent study concludes that with ordinary caution even the highest power lasers in this range (1-5 mW) are safe. Over 90% of the exercises described for the student laboratory, from elementary school science to undergraduate physics level, can be performed with low-power helium-neon gas lasers in the 0.5-1.0 mW range. The Metrologic ML-669 modulatable student laser used in most exercises covered in these pages has a nominal 0.8 mW visible output which places the device within the Class II category, as defined by the Department of Health, Education and Welfare for laser safety. To insure absolute safety, however, the following precautions and safety procedures are recommended:

1) Treat all laser beams with great respect.

2) Never look into the laser window (even when turned off) or stare into the beam (on axis) with either the naked eye or through binoculars or a telescope at a distance.

3) Do not rely on sunglasses as eye-protecting devices.

4) Never point the laser beam near anyone else's eyes.

5) Cover all windows to protect passers-by.

6) Never leave lasers unattended while activated. If not in use, disconnect A.C. power cable.

7) Room illumination in the work area should be as high as is practicable to keep the eye pupil small and reduce the possibility of retinal damage due to accidental exposure.

8) Remove all superfluous and highly reflective objects from the beam's path. These include rings, watches, metallic watchbands, shiny tools, glassware, etc.

9) Beware of electrical hazards: Ungrounded frames or chassis' and inadequately insulated power cables. Adequate grounding should be provided for the laser case and the laser should never be operated without a protective cover.

10) Never attempt any adjustments to the laser plasma tube or associated electronics with the laser plugged in. First, disconnect the power

cable and then discharge capacitors. Lethal current levels at high voltage is present inside the laser chassis.

11)   While operating outdoors (laser communications, speed of light experiments, etc.) do not point the laser at passers-by and do not track vehicular or airborne traffic with the laser beam.

12)   Do not operate the laser in rain, snow, fog or heavy dust. Potentially dangerous, uncontrolled specular reflections can result.

References:

Lloyd, L.B., Popular Electronics p41 (Dec '69)
Myers, G.E., Electro-optical Systems Design p30 (Jly '73)
Tinker, R., Physics Teacher v11 p455 (1973)
Weichel, H., Danne, W.A. and Pedrotti, L.S., American Journal of Physics v42 p1006 (Nov '74)
Federal Register v40 n148 Pt. II (31 Jly '75)

Care and Maintenance of Equipment

The following suggestions for routine care of the equipment will help to prolong its useful life:

1)   To save the plasme tube, the laser should be shut off and disconnected when not in use.

2)   To prevent dust from entering the laser housing, the laser should be stored in a dry room and covered with plastic or other suitable covering.

3)   If possible, all chemicals should be kept in a separate store-room, away from the laser and its accessories.

4)   Before and after each use wipe smudges and fingerprints from lenses, prisms and other accessories with a soft cloth, or, preferably with a good quality lens tissue.

5)   After each use, wrap lenses, prisms, glass plates, filters and all fragile accessories in lens paper or soft tissue and place them in individually marked envelopes or boxes. In this way they may be easily located and safely stored.

6)   Never use solvents to clean plastic parts, polarizing and color filters, photographic films such as holograms, diffraction grating replicas and similar items.

The Optical Bench and Basic Accessories

    The basic components needed to perform the exercises and lecture demonstrations described in this book are listed below.

Name of Item:	Description:
Laser... (Metrologic ML-968)	0.8 mW student modulatable laser; TEM_{00} mode; 15% modulation from 50 Hz to 500 kHz range; beam diameter 1.2 mm; beam divergence 1.0 mRad; random polarization.
Optical Bench...	The optical bench is made of heavy-duty aluminum (or iron casting) of triangular cross section. The benches are grooved on both sides to engage the screw which clamps the laser carrier and various component carriers. 1/4, 1/2 and 1 meter long units.
Laser Carrier... (Adjustable model)	The laser carrier is an adjustable platform which firmly affixes the laser to the optical bench. The clamping screw locks the carrier in position on the bench. An adjustment screw at each end of the carrier provides for precise positioning of the laser beam. The laser beam may be moved up, down, and to either side.
Component Carrier...	The component carrier has a clamping screw at its bottom which firmly attaches it to the optical bench. Another screw is at the top and is used to mount lenses, plates, filters, slides and other components in the path of the laser beam. Component carriers come in units of varying base length and height.
Pin Carrier...	These are component carriers designed for holding 13.7 mm pins of various lengths.
Mounting Pin...	Pins are used to support certain accessories on carriers. The standard pin diameter is 13.7 mm; the length 6 cm and 12.5 cm. The pins have a 1/4" x 20 female thread at one end and are supplied with a 1/4" x 20 headless screw. The pin may therefore take components with either male or female threads. Mounting pins may be adjusted up and down on the pin carrier. Often small tables are attached to the pin to carry a prism or transparent vessel, or a 35-mm camera.

Circular Component Holder...

The circular component holder is used to mount lenses, prisms, filters and other components. Three cushioned screws allow precision positioning in the optical path. This holder will hold components from 6 mm (1/4") up to 50 mm (2") in diameter.

Spatial Filter...
(Metrologic 60-618)

The lens-pinhole filter provides a clean, noise-free laser beam.

Photometer...
(Metrologic 60-230)

The photometer is a battery-operated instrument. It contains a silicon light sensor, a solid state amplifier, and a sensitive micro-ammeter. An output jack is provided for connection to a strip chart recorder, an external projection meter, or an audio amplifier system.

Glass Plates...

Microscope slides, photographic cover glasses, and good optical quality glass blocks (plates) of various thicknesses.

Front Surface Mirrors...

Mirrors with the silvering on the front surface. Various sizes.

Glass Plate Beamsplitter...

The glass plate beamsplitter held in the laser beam at an acute angle permits part of the light to be transmitted and part to be reflected, thus producing two laser beams.

Cube Beamsplitter...
(Edmund No. 30,329)

Two right angle prisms, one of which carries a thin metallic coating, are cemented together. It gives good results over large angles of incidence while eliminating "ghost images" present when working with glass plate beamsplitters.

Prism Set...

Several 90-degree and 60-degree prisms of various sizes.

Lens Set...

Assortment of double-convex and plano-convex lenses from 25 mm to 250 mm focal length. Double-concave and plane-concave lenses of -5 mm and -25mm focal length. Various diameters. Short length of 2-5 mm diameter plexiglas (lucite) rod to be used as a cylinder lens.

Set of Optical Models...
(Edmund No. 71,181)

Used for lazy susan exercises. A 7-piece prism/lens set. Double-convex lens, double-concave lens, semi-circular disc, two 90-deg. prisms, Dove prism, and 60-deg. prism made of plexiglas.

Polarizing Filters...
(Set of 2)

Mounted in such a way that they can be held rigidly in component carriers.

Color Filter Set...

The red, green, yellow and blue color filters (glass or gelatin) are mounted for insertion in the component carriers.

Diffraction Grating Replica...
(Edmund No. 40,272)

Card mounted transmission type gratings; 13,400 grooves per inch.

Miscellaneous...

Lens cleaning supplies; white cardboard viewing screen; ground glass screen; graph paper with fine grid; glass vessels such as fish tanks and smaller units; razor blades; thin wire; ceramic magnets (flat and ring shaped); putty (non-hardening); commercially available holograms.

"""""""""""""

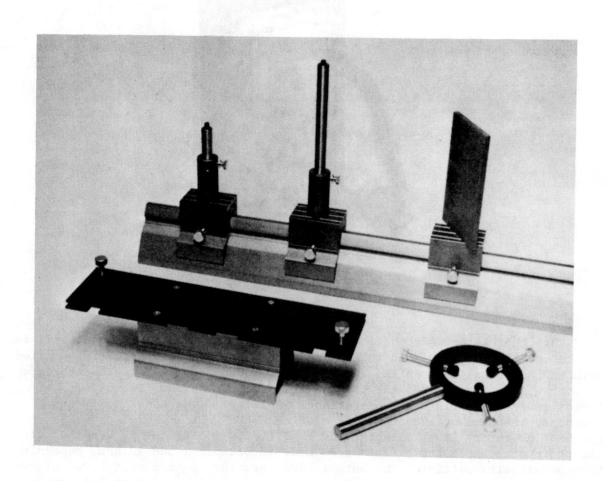

Fig. 6. Optical bench, pin carriers, laser carrier and circular component holder

Laser Power Meters and Detectors

Most photodetectors and light meters designed either for photographic use or in physical science laboratories, will respond to a laser beam and may be used to give indications of relative brightness of the beam in direct current and low frequency modulation applications. However, most of these units are unsatisfactory for quantitative work with lasers because of range limitations, aperture size, and other design factors.

Where a reasonably accurate laser power meter was required, and cost was a factor, the 60-230 Metrologic Calibrated Photometer was used for many of the exercises described in this book.

Fig. 7. The 60-230 Photometer

This photometer is a photocell driving a microammeter. The photocell is a 1 cm^2 silicon cell mounted in a 25 mm cylinder. Each photocell is supplied with an individual calibration curve. The range is from 0.3 microwatts/cm^2 to over 0.3 watts/cm^2 when the pinhole cell cover is used. Two 9-volt batteries power the solid state operational amplifier. The sensitivity and frequency response of the amplifier also make the unit a useful general purpose laboratory light detector and amplifier. The frequency response of the photometer amplifier extends from D.C. to between 1 and 10 kHz, depending upon sensitivity setting. The output plugs provide up to 5 volt D.C. intended for large projection display meters. The diagram (Fig. 8) shows the system.

An oscilloscope (such as Heath/Schlumberger Model 50-107A) may be attached to the output terminals of the photometer to observe signals from D.C. to up to 10kHz. The photometer can be used, for example, with an interferometer to measure either slow fringe variations or fringe counts up to 10,000 fringes per second (such as Heath/Schlumberger Model SM-118A counter).

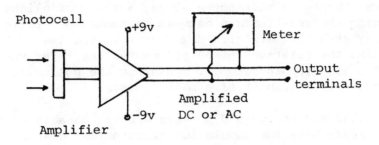

Fig. 8. Block diagram: 60-230 photometer

In the study of diffraction patterns from various slits, apertures and gratings, it is helpful to have a permanent record of a given pattern. The photometer/amplifier can be used to measure the light intensity at various points in a diffraction or interference pattern. The output of the photometer is connected to a strip-chart recorder (such as the Heath/Schlumberger Model SR-201A). The photometer is moved laterally across the diffraction pattern at a slow, uniform rate. When the entire width of the pattern has been recorded, the result is a finished graph of the intensity levels of the diffraction pattern.

The Metrologic 60-530 Professional Laser Power Meter has the general features of the 60-230 Photometer plus greater accuracy, a much more sensitive photocell and amplifier system, and a larger meter. The user may read down to one nanowatt on the 60-530. This unit is shown in Fig. 9.

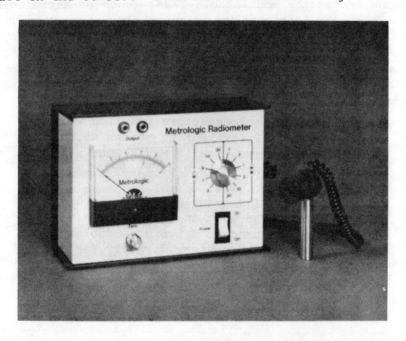

Fig. 9. The 60-530 Radiometer

A precision photometer is necessary for measuring many quantitative phenomena. A few examples: measuring the radiometric intensity from a He-Ne laser... plotting beam intensity distribution... finding the width of a laser beam between half-power points... measuring the reflected power from various surfaces at various incidence angles... confirming Lambert's cosine law... measuring the transmitted power through a microscope slide; a stack of slides; through colored filters... finding the relationship between transmission, reflection and absorption ... measuring Rayleigh scattering... demonstrating the law of Malus... ellipsometry... measuring the relative intensity between the reference beam and the subject beam in a holographic setup... measuring the pollution content (turbidity) of water, and many other quantitative measurements.

Both the 60-230 and the 60-530 photometers are capable of demodulating an audio-modulated laser beam for modulation frequencies between 0 and 10 kHz. They can, therefore, be used as the receiver in a laser voice communications link. Figure 10 depicts such a link.

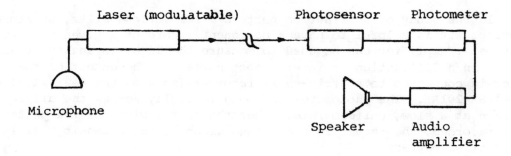

Fig. 10. Laser communication link

A microphone drives the modulatable laser (such as Metrologic ML-669 or ML-939). The laser beam is picked up by the photo-sensor/amplifier (such as Metrologic 60-230 or 60-530). The output is further amplified by feeding it into an audio power-amplifier/speaker combination. For voice communications over longer distances the laser beam needs to be collimated. The receiver end of the laser beam path should have a large Fresnel lens to focus the incoming beam on the photosensor.

While the above described photodetectors function well between 0 and 10kHz, there are applications where an instrument is needed which responds to light variations of from 200 Hz to an excess of 2 MHz. (Such a low-cost unit is the Metrologic 60-255 Photodetector shown in Fig. 11.

Two outputs are provided. The least sensitive is 0.6 volts/mW and the most sensitive is 6 volts/mW. The unit is powered by a 9-volt transistor battery. The 60-255 Photodetector may be used to drive an audio amplifier for voice communications, or to drive a video monitor directly for TV transmission on a laser beam. It can also serve as the receiver in the speed of light experiment.

Fig. 11. The 60-255 high speed photodetector

Exploratory Exercises

A) <u>Color.</u> The helium-neon laser generates an intense output of red light
at the wavelength of 632.8 nanometers (nm). In addition to the
red laser light, the helium of the laser produces some incoherent blue
and green light.

(A-1) Project the laser beam on to a sheet of white paper held about
25 centimeters in front of the laser. Then, in a darkened room,
observe the colors on the paper.

(A-2) One at a time, insert red, yellow, green and blue cellophane
filters into the path of the laser beam. Observe then, for each
filter, the resulting colors on the paper screen.

(A-3) Repeat (A-2), first with a single thickness of a given colored
cellophane, then with two thicknesses (two sheets), then three
thicknesses, and so on. Observe how the absorption of light
depends upon the thickness of the absorber.

(A-4) Try combinations of two or more different filters at the same
time; try to predict the resulting colors in advance.

B)   Scattering.   Objects are visible only because of the light they scatter
                     toward the viewers' eyes. Laser light, like any other, is
invisible unless minute, air-borne particles, such as those of dust or
smoke, are in its path to reflect and scatter it. It is this scattering
of the light which makes it possible to "see" the laser beam. To illustrate this concept try the following exercises:

(B-1)   In a darkened room, turn the laser on and aim the beam so it
         traverses a path 90 degrees to the view of observers. If the air
         is "clean" the laser beam will be barely visible, if at all. To
         make it more visible some air-borne particles will have to be put
         in its path. For this purpose use the discharge from any air-
         freshener type aerosol spray, or even water sprayed from a per-
         fume atomizer. With the spray in the beam's path the beam should
         become clearly visible and remain so for as long as the sprayed
         particles remain suspended in the air. The finer the spray, the
         longer the suspension time. The same effect can be achieved with
         chalk dust from a blackboard eraser or with the particles that
         make up any kind of smoke.

(B-2)   The Smoke Tank is an aquarium type tank of 2 to 10 gallons capac-
         ity and fitted with a removable top. When the tank is filled with
         smoke, the light scattered by the particles will make the laser
         beam path visible. A good smoke generator is an incense cone.
         Place the cone in the tank and light it. Blow the flame out after
         a minute and immediately cover the tank with its fitted top.
         Lenses, mirrors and prisms may be placed inside the tank so that
         light paths can be seen before and after reflection or refraction.
         If no incense cone is available, a small piece of burning string
         or rope will do as well.

(B-3)   The Smoke Tank used in exercise (B-2) is now used as a Water Tank.
         First, direct the laser beam so that it passes through the display
         tank filled with boiled or distilled water. The path of the beam
         will probably not be visible in the water. The addition of a
         small amount of scattering particles to the water will make the
         beam visible...
         * A few drops of skim (non-fat) milk mixed with the water.
         * A pinch of non-fat dry milk (powdered milk). Stir thoroughly.
         * Sufficient sodium fluorescein to give the water a pronounced
           color.
         * Hypo, the photographic fixing salt dissolved in the water plus
           a few drops of strong sulphuric acid. This causes sulphur
           particles to precipitate slowly.
         * A few drops of liquid detergent added to the water. Stir until
           uniformly mixed.
         * A drop of polystyrene latex microspheres (used in Millikan
           electron charge apparatus and for electron microscope calibra-
           tion) in the water. The fine particles will remain suspended
           indefinitely in the water. (Microspheres are available from
           Duke Standards Co., Palo Alto, California).

         The Water Tank is excellent for vivid demonstrations of (a) re-
         flection and refraction effects; (b) the critical angle;
         (c) total internal reflection. (See P. DiLavore, Phys Teacher,
         p247, Apr. '75).

(C)   The Spreading of the Laser Beam.   A laser produces a narrow beam of light. Since perfect parallelism is an unobtainable goal, each laser produces a beam that is somewhere between perfect parallelism and a certain amount of spread. The amount of angular spread (also called "beam divergence") is listed by the manufacturer among the specifications given for the laser.

The approximate value of the angular beam spread is $\Delta\theta \approx \lambda/D$, where λ is the wavelength of the light and D is the diameter of the beam as it exits from the plasma tube. Let us say the laser beam has a diameter of 1 mm at the source and the wavelength is 633 nanometers. Question is: How much does the beam diameter increase in a distance of 50 ft.? The angular spread of the beam is

$$\Delta\theta \approx \lambda/D \approx (6.33 \times 10^{-5} \text{ cm})/(0.1 \text{ cm}) \approx 6.33 \times 10^{-4} \text{ rad.}$$

The angular spread times the distance L = 50 ft (approximately 1500 cm) gives the spatial spread of $W \approx L\Delta\theta = L\lambda/D \approx (1500).(6.33 \times 10^{-4})$, approx. 0.95 cm (almost 10 mm). This can be nicely demonstrated in a large auditorium or a long corridor.

1.   Measure the diameter of the laser beam by holding a sheet of mm graph paper in front of the laser. Record the diameter of the bright red spot on the graph paper.

2.   At a distance of 15 meters in front of the laser, measure the diameter of the beam again with the help of the graph paper. Compare the measured value with the calculated one. For this calculation use the values given by the manufacturer for beam diamater and angular beam spread (beam divergence).

3.   Repeat the exercise outdoors at night for long-distance measurements. NOTE: The laser beam is brightest at its center and intensity falls off toward its outer edges. It becomes difficult, therefore , to determine exactly where the beam ends. For consistency, the half-power points are used for the beam's edges. At these points a photometer will indicate the beam to be only one half as intense as in its center.

4.   Make a graph plotting the beam diameter versus distance.

5.   Calculate the angular beam spread by dividing the change in beam diameter by the distance from the laser. The quotient will be the angle measured in radians. Multiply the value by 1,000 to obtain the angular beam spread (beam divergence angle) in mRad (milliradians).

D)   Beam Intensity.   Compared with a conventional light source one of the advantages of a laser is that the beam is very intense and that its brightness does not fall off rapidly with distance.

(D-1)   Hold a white sheet of paper in front of the laser and walk away from the laser fifty meters or more while continuously observing the intensity of the bright red spot. NOTE: Never look directly into the laser beam yourself!

(D-2)   Measure the light intensity at increasing distances from the laser using a photometer. Make sure that the opening which admits light into the photometer is the same diameter as the laser beam at its exit window, its narrowest portion.

(D-3)   The raw, undiverged beam from a low-power He-Ne laser can easily be seen near the center of a 25-watt frosted incandescent bulb. If a 1 mW laser is used, than the output power of the light bulb is 25,000 times that of the laser. If only 5% of the bulb's output is in the visible part of the spectrum, the ratio is 1,250:1. Assuming that the frosted glass reflects less than 10% of the incident laser beam this raises the ratio to a minimum of 12,500:1. The fact is that even a 0.5 mW laser emits a beam intense enough to be seen on a 25 W frosted light bulb: - a ratio of 25,000:1. Thus, we see that, typically, a small laser source is better than conventional sources by a factor of 10^3 to 10^5 in brightness. This quality of brightness from the laser makes it unique for use in optics demonstrations. But also, it means that one must not look directly into the laser beam and precautions must be taken that the beam, or its reflection, does not shine directly into the eyes of others in the vicinity.

""""""""""""

Distance Measurement by Triangulation

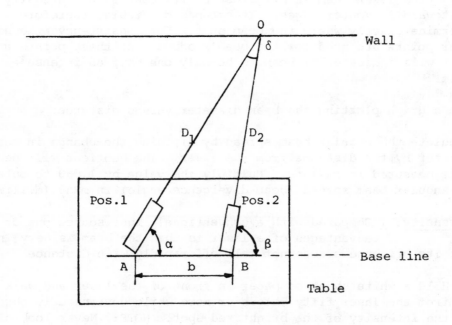

Fig. 12. Laser triangulation

The laser is set up on a sturdy table. The beam is directed at a distant object or at a spot O marked on a wall.

With the protractor the angle α between the side of the laser (the laser beam direction) and the base line in Position 1 is measured. Next, the laser is moved along the base line to Position 2 and again the laser beam is aimed to strike the spot O on the wall. It is important to make sura that the back of the laser is along the base line in Position 2. The angle β is now measured in Position 2. Finally, the distance b (the translation of the laser between Positions 1 and 2 is measured.

From these three measurements: two angles and a distance (b) which is the short side of triangle AOB we can determine the distances D_1 and D_2 from the laser locations A and B to the distant point O on the wall.

$$D_1 = b \; \frac{\sin \; (180-\beta)}{\sin \; (\beta-\alpha)}$$

$$D_2 = b \; \frac{\sin \; \alpha}{\sin \; (\beta-\alpha)}$$

" " " " " " " " " "

Laser Alignment — the X-Y Position Indicator

A laser is the best and simplest device for drawing a straight line through space. He-Ne lasers produce narrow, straight beams that can be projected for hundreds of meters. These lasers operate continuously and their red beams can be used as guidelines for alignment jobs. The laser tubes are carefully centered within the stainless steel housings so that the beams are pre-aligned and parallel with the housings. Thus, by aligning the steel cylinders mechanically, the laser beams are aligned optically. These encapsulated alignment lasers are low in cost, light in weight, can withstand rough handling and are operated on either battery power or standard 110/220 volt electrical power. Just a few of their uses: tunnel alignment, bridge construction, masonry work, trench alignment and pipe laying. Even small construction companies and subcontractors such as plumbers, electricians and surveyors employ alignment lasers as time, cost and labor-saving devices.

Not only can a straight line be defined but "squareness" and "parallelism" as well. For straight-edge applications the laser can be used alone.

The alignment accuracy obtainable depends on the ability to define the center of the laser beam. The beam stability is another important factor. Once the laser is set up it must be capable of providing a beam which continues, without disturbance, to point in the initially determined direction.

In some applications, such as tunnel boring, a crude method of estimating the beam center suffices. A graticule of concentric rings is placed in the path of the beam and the machine operator guides the machine so as to keep the beam approximately on the center of the graticule.

An enhanced method of alignment is provided by the use of a coarse circular diffraction grating. When the laser beam is incident normally on a circular diffraction grating, a circular fringe pattern is produced in any plane parallel to the grating on the far side of the grating. The center of this fringe pattern defines a straight line which can be used for continuous alignment in conjunction with a graticule of concentric circles.

For more precise work an electronic detecting system must be used for the accurate sensing of the center of a laser beam. The Metrologic Position Indicator (Model 60-228) may be used with any TEM_{OO} laser at distances up to 30 meters with a precision of approximately 0.1 mm. When the laser is used in conjunction with a collimator (such as the Metrologic 60-223), the Quadrant Detector is accurate at distances up to 200 meters.

The Quadrant Detector is the segmented photo-sensitive cell itself. It is mounted in a rugged, independent housing, connected to the meter)called "Position Indicator"). The Quadrant Detector may be mounted on an optical bench or by custom mounts designed for particular applications.

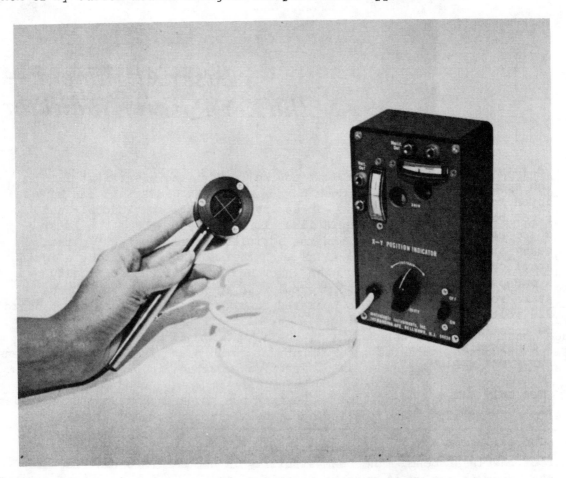

Fig. 13. Position Indicator 60-228

The Quadrant Detector consists of four symmetrically placed silicon photo-cells. Diametrically opposed cells are linked so as to produce a null reading when each is illuminated by equal intensities. The beam is centered when two

20

null readings from each diametrically opposed pair are obtained simultaneously.

The laser beam is directed so that it strikes the face of the photocell as nearly perpendicular to the cell as possible. The deflection of the X and Y meters indicates the position of the laser beam on the photocell.

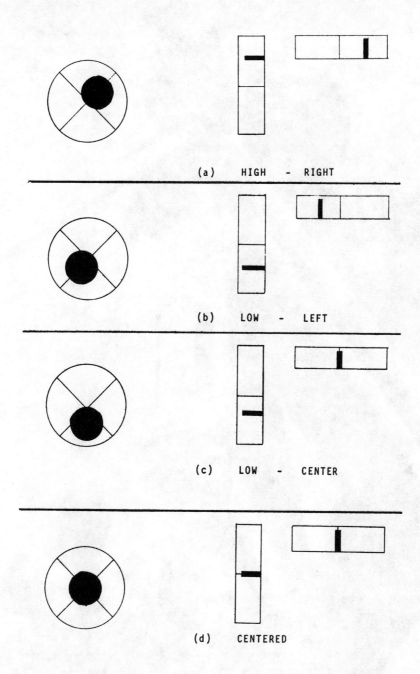

Fig. 14.   The Quadrant Detector readout

If the detector is intended to locate the center of a fixed laser beam, the horizontal and vertical position of the photocell is slowly adjusted until both meters read "zero" (mid scale). Conversely, if the laser beam is to be aligned with the fixed detector, the position of the laser beam is adjusted until both meters read "zero."

Fig. 15. MODERN TRANSIT incorporates a Metrologic Instruments' encapsulated laser (cylinder at bottom) to help surveyors speed alignment work.

Lazy Susan Optics Table

Ray diagrams are commonly used to illustrate light paths in optical devices. The illuminating system of the lazy susan optics table allows the observer to see <u>actual</u> paths created by <u>real</u> beams of light as they are reflected and refracted by mirrors, glass blocks, prisms, lenses and combinations of these optical elements.

To produce ray streaks on the optics table, the narrow, parallel laser beam is substantially diverged by a cylindrical lens which has an extremely short focal length. The power of a cylindrical lens to converge light rays when its axis is in the horizontal position is in the vertical plane only, no refraction is produced in the horizontal plane (see Fig. 16).

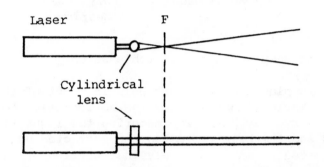

(a) Setup viewed from the side: a fan of rays emerges from the cylindrical lens.

(b) Setup viewed from the top.

Fig. 16. Cylindrical lens creates a thin sheet of light

A glass, lucite, or plexiglas rod is used as a cylindrical lens. It has to have a diameter sufficiently small to produce an extremely short focal length but it has to be slightly greater than the nominal width of the raw laser beam. A diameter for the cylindrical lens that gives acceptable results is 2-4 mm. A simple way to mount the cylindrical lens is to cement the plexiglas rod (1-2 cm long) on a doughnut shaped magnet of approximately 20 mm outside diameter and a 6-8 mm center hole. The magnet holds the cylindrical lens to the laser housing where the beam emerges. This mounting allows fine adjustment of the plexiglad rod relative to the beam. As shown in Fig. 16, the axis of the plexiglas rod must be oriented horizontally and the laser beam centered on its cylindrical surface. If the laser is now aimed at a wall or screen, a narrow, bright, vertical line will be seen.

Note: A single thin sheet of light from the laser can be used for certain demonstration purposes, but, to illustrate many phenomena, more than one light ribbon is necessary. Instead of one light ribbon, one uses several parallel sheets of light. This can be accomplished with the help of beam multipliers, to be described later in the book.

The optical models (glass slabs, mirrors, prisms and lenses) rest on a platform that rotates like a lazy susan. The smallest 3-inch "Lazy Susan" bearing (Edmund Scientific, Cat. No. 40,600) is satisfactory for making turntables up to 18" diameter. Bearings are also available in 4", 6" and 12" sizes. The bearing is first mounted on a plywood base with wood screws. The turntable is a heavy, white cardboard, such as used for mounting photographic enlargements. This board can be attached to the top-side of the bearing with common adhesive. The optical models used are part of a 7-piece prism/lens demonstration set (Edmund Scientific, Cat. No. 71,181).

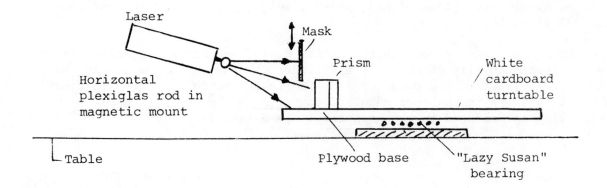

Fig. 17. Side view of lazy susan optics table setup

As seen in Fig. 17, a prism is placed on the turntable in front of the laser, with the laser aimed at an angle slightly below the horizontal. Looking down at the turntable, one can observe ray paths representing the light ribbon originating at the cylinder lens and the beam reflected and refracted by the prism. The vertical mask is adjusted to make sure only that portion of the light ribbon reaches the turntable which passes through the optical model (prism in this case) and those above it are blocked.

NOTE: While it is advisable to use lasers with caution at all times, the hazards associated with the lazy susan setup are minimal. Assuming the maximum power rating of the laser beam is 1 mW, the beam width 1 mm, and a 2-4 mm diameter cylindrical lens is used. The calculated intensity at 1 meter that could enter the eye is approximately 100 microwatts. The cylindrical lens reduces this value by beam divergence so that a further reduction by a factor of 100 is achieved in intensity that could enter the eye.

Reference:

Hughes, D.B., Karzmark, C.J., and Rust, D.C., Phys Med Biol v18 p881 (1973)

Lazy Susan Exercises

1. <u>Reflectance and Transmittance</u>.

    NOTE: All drawings show the setup from above unless otherwise noted.

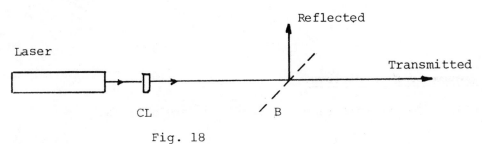

Fig. 18

(1-A) A thin sheet of laser light, originating at the cylindrical lens (CL)
       is incident obliquely on a beamsplitter, B. To split the light, a
       thin photographic slide cover glass or a microscope slide can be used.
       Part of the light is transmitted and part is reflected by the beam-
       splitter. The intensity of the transmitted and reflected rays may be
       observed on the lazy susan table or may be measured beyond the
       boundaries of the table with a photometer (see Fig. 18).

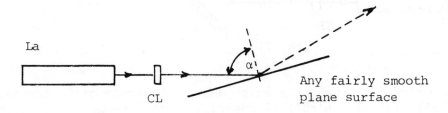

Fig. 19

(1-B) A frosted piece of glass is placed on the lazy susan table in the path
       of the light. Cover the back (smooth) surface with black matte lacquer.
       At first, the frosted surface is directed toward the incident light:
       the reflected light is scattered. As the frosted glass is rotated to
       increase the angle of incidence, the reflected light becomes more
       sharply defined and with glancing incidence (i.e., when the angle of
       incidence, α, is close to a right angle), a mirror like reflection of
       the laser light is obtained. Substitute a piece of smooth paper (small
       filing card), or the undeveloped emulsion side of a photographic plate
       for the frosted glass and repeat the exercise (see Fig. 19).

(1-C) About 4% of the light incident normally on an air-glass interface will
       be reflected back. The exercise is performed with the help of a package
       of microscope slides or photographic slide cover glasses. If the laser
       beam is directed into a stack of these glass slides most of the light

will be reflected (see Fig. 20).

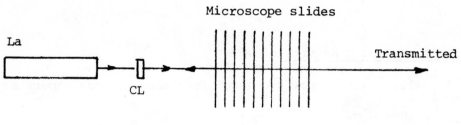

Fig. 20

2. Exercises with Mirrors.

NOTE: The mirrors used are front surface mirrors.

(2-A) The Law of Reflection. An incident sheet of laser light is projected obliquely to the mirror surface producing a reflected ray on the lazy susan table. It is seen that the angle of incidence (α) equals the angle of reflection (β) as shown in Fig. 21.

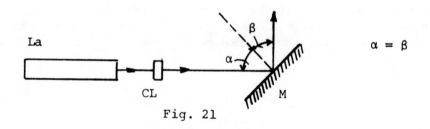

Fig. 21

If the mirror is then rotated to change the angle of incidence, the same relationship ($\alpha'=\beta'$) is observed in the second position (Fig. 22).

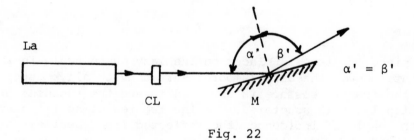

Fig. 22

(2-B) If the angle of the mirror rotation (θ) between Positions #1 and #2 is measured, it is found that the reflected ray is deflected by 2θ. (see Fig. 23).

(2-C) Virtual Image in a Plane Mirror. After reflection at mirror M_2, all rays (two rays in this exercise) originally diverging from the object point P now diverge from the image point P'. Point P' is called the virtual image of P. The sheet

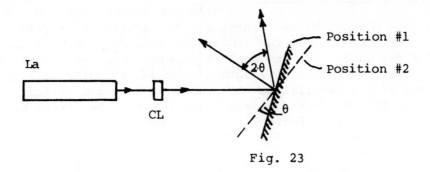

Fig. 23

of laser light is split in two with the beamsplitter, B. The small
front surface mirror, M_1 re-directs ray #2 to cross ray #1 at object
point P. Extensions of the rays reflected by the larger mirror, M_2 are
drawn behind the mirror until they intersect to locate virtual image
point, P'. (see Fig. 24).

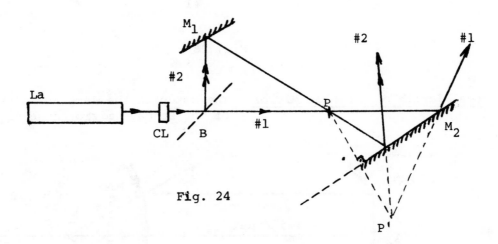

Fig. 24

(2-D) In this exercise two plane mirrors are at an angle to each other. The
arrangement is such that ray A is deviated through a constant angle (β)
equal to twice the angle α between the two plane mirrors, M_1 and M_2.
(Note that if $\alpha = 45°$ then $\beta = 90°$ and we have the mirror equivalent of
a penta prism.) Slowly rotate the lazy susan table and observe that
the relationship $\beta = 2\alpha$ is valid. (see Fig. 25)

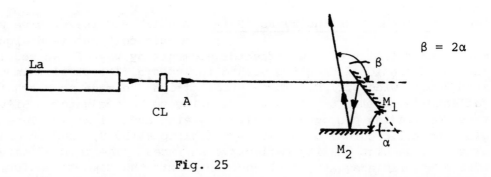

Fig. 25

(2-E) <u>Focal Length of a Concave Mirror.</u> The sheet of light is split in two
with the beamsplitter, B. The small
front surface mirror, M$_1$ re-directs ray #2 parallel to ray #1. The
center of curvature, C, of the concave mirror, M$_2$, can be located by
means of a ray which reflects back on itself. This ray, r, is indicated
with dotted line. The focal length of the concave mirror: f = r/2.
Rays #1 and #2 are parallel to the principal axis and reflect through
F (principal focus) located midway between C and the vertex of the
concave mirror on the principal axis (see Fig. 26).

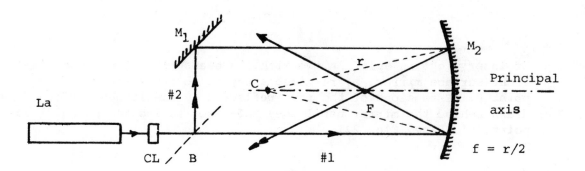

Fig. 26

(2-F) <u>Focal length of a Convex Mirror.</u> The setup is the same as in the case
of the concave mirror (see Fig. 27).

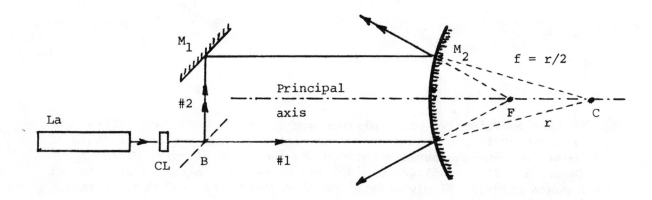

Fig. 27

(2-G) <u>Two-dimensional Modeling by Ray Paths.</u> A quick and inexpensive way to
obtain comparative design infor-
mation on a family of 3-D reflector elements by way of 2-D modeling.
The two-dimensional modeling method includes a flat steel shim stock cut
into 2-5 cm wide strips. These steel strips are made to match the
reflector's profile on the design/drawing to be evaluated. Silver
tape for binding photographic slides (Leitz Special Silver Tape) or hi-
gloss metallic chrome on mylar tape (Chartpak #5019) used in graphic
arts provide high quality reflective surfaces. The thin adhesive substrate affords adequate compliance to permit the tape to conform to
the shim stock through a range of curvatures. A small mirror, RM,

prisms of 45-45-90 degrees as totally reflecting surfaces in many optical instruments.

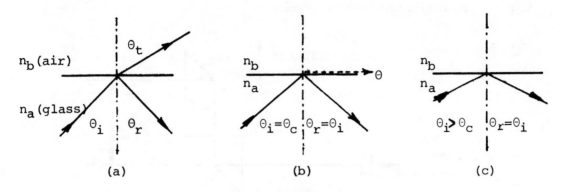

Fig. 32

(4-A)   The setup shown in Fig. 33 can be used to demonstrate total internal reflection and the critical angle. A diffraction grating or other beamsplitting means (described elsewhere in the book) are used to generate multiple ray-streaks on the lazy susan table.

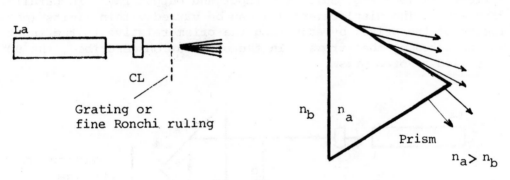

Fig. 33

5.   <u>Exercises with Prisms.</u>

(5-A)   <u>The Right-Angle (90°) Prism - Total Internal Reflection.</u>(One reflection)

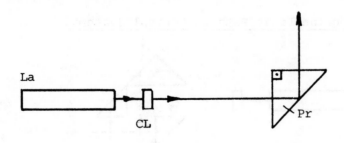

Fig. 34

The right-angle prism deviates rays by 90° through a single internal reflection (see Fig. 34).

(5-B) When using two 90° prisms in the arrangement shown in Fig. 35, the ray
leaving the second prism (Pr$_2$) is laterally displaced relative to the
original direction. The displacement (d) can be controlled by adjusting Pr$_2$ as indicated by the arrows.

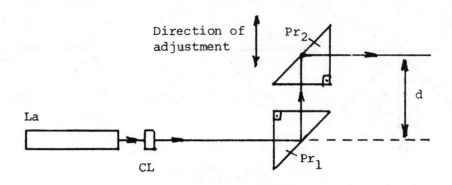

Fig. 35

(5-C) <u>The Right-Angle (90°) Prism - Total Internal Reflection.</u>(Two reflections)

When the 90° prism is turned so that the ray enters the hypotenuse at
right angles, a total deviation of 180° is achieved by two internal
reflections (see Fig. 36). The input and output rays are parallel and
displaced. The displacement (d) can be varied within limits (physical
limits of the prism) by adjusting the prism relative to the input ray
as indicated by the arrows. In the orientation described, the 90° prism
is called a Porro prism.

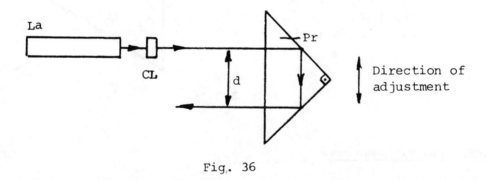

Fig. 36

(5-D) <u>Means for Folding the Light Path in Optical Systems.</u>

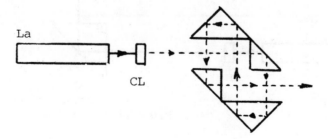

Fig. 37

In many optical systems such as telescopes, binoculars, and gunsights, it is desirable to have a long optical path without the inconvenience of unnecessary overall length. Fig. 37 shows a simple arrangement of four right-angle or Porro prisms which leaves considerable space in the optical path for the introduction of optical elements such as lenses.

(5-E) <u>The Reversing (Dove) Prism.</u> A right-angle prism is used almost exclusively in collimated light. Refraction and total internal reflection cooperate to reverse the positions of two parallel rays. The truncated version of the 90° prism (to reduce size and weight) is called the Dove prism (see Fig. 38).

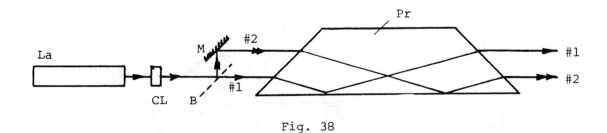

Fig. 38

(5-F) <u>Angle of Minimum Deviation.</u> The deviation suffered by a monochromatic beam on traversing a given prism is a function only of the incident angle at the first face. The smallest value of δ is known as the minimum deviation δ_m and it is of great practical interest for determining the refractive index of a transparent substance.

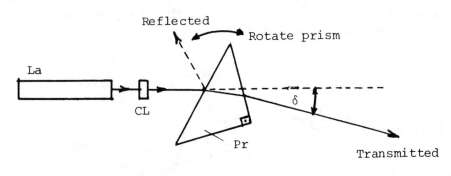

Fig. 39

When a 90° prism is used, the deviation of a sheet of light is most easily observed on the lazy susan table when the light passes through the prism just below one of the 45° angles (see Fig. 39).

Repeat the exercise with an equilateral (60-60-60) prism. The exercise using the equilateral prism duplicates the conditions present in adjustable spectroscope prisms. The deviation (δ) assumes a minimum value with the light passing through the prism symmetrically (Fig. 40).

In case of minimum deviation (δ_m) the refracted beam (A') and the beam reflected from the base of the prism (B') run parallel (see Fig. 41).

When a plane mirror is parallel to the base of a prism at minimum

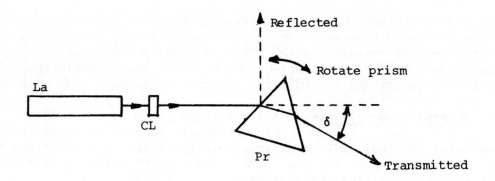

Fig. 40

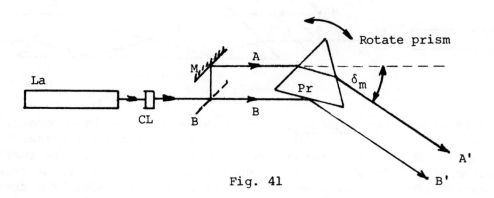

Fig. 41

deviation as shown, the light leaving the plane mirror is parallel to that entering the prism (see Fig. 42).

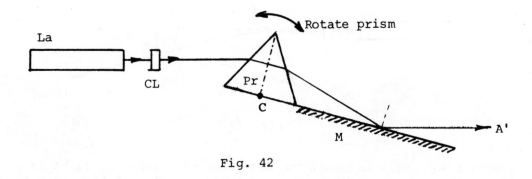

Fig. 42

C is the center of rotation, M is a front surface mirror.

(5-G) The Refractive Power of a Prism is a measurement of the distance the refracted ray deviates from the path of the incident ray at one meter from the prism (see Fig. 43).

The unit of refractive power of a prism is the centrad (1/100 radian or diopter). A prism of one centrad (one diopter) bends light to such an extent that when a refracted ray travels one meter beyond the prism it deviates a distance of 1 cm from the path of the incident ray. For

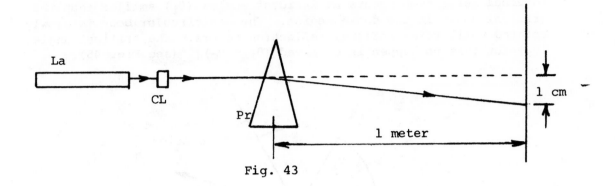

Fig. 43

example, a three centrad prism will deflect light by 18 cm at a distance of 6 meters from the prism.

6. Exercises with a Semicircular Body.

One of the most convenient ways to demonstrate the laws of reflection and refraction and to measure the refractive index is by using a semicircular body placed concentrically on the axis of the lazy susan table.

(6-A) **Light Incident at Center of Disc.** By turning the lazy susan table to change the angle of incidence, the angles of incidence and refraction can be measured to determine the index of refraction. The three basic laws of reflection and refraction can be demonstrated: 1) The incident, reflected and refracted rays all lie in the plane of incidence; 2) $\theta_i = \theta_r$; and 3) $n_i \sin \theta_i = n_t \sin \theta_t$. The light emerges perpendicular to the tangent at the point of exit and thus will not undergo a second deviation (see Fig. 44).

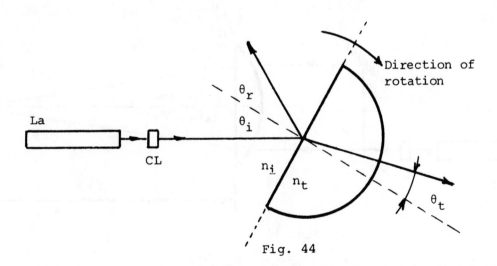

Fig. 44

(6-B) **Light Incident at Right Angles to Tangent - The Critical Angle.**

In this arrangement the light will always enter the cylindrical surface of the disc radially. It will thus be parallel to the normal of the glass and not be deflected. As the light goes from an optically dense to an optically light medium at the plane surface, an angular deflection (θ_t) of light will occur away from the normal. Partial internal reflection occurs at incident angles (θ_i) smaller than the critical angle in the dense medium. The semicircular body is slowly rotated until total internal reflection occurs. The critical angle (θ_c) for this to happen is observed ($\theta_i = \theta_c$), (see Fig. 45).

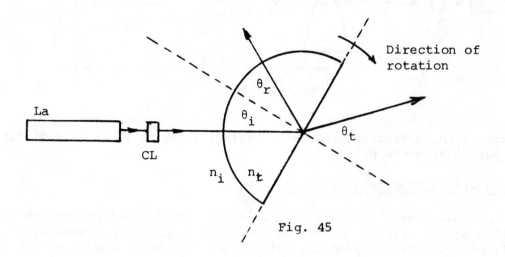

Fig. 45

7. <u>Exercises with Lenses.</u>

(7-A) <u>Focal Length of a Positive (Convex) Lens.</u> With the help of the beam-splitter (B) and a small front surface mirror (M), two parallel ray streaks are projected upon the lazy susan table. Care must be taken in establishing parallelism. The plastic model of a positive lens (plano-convex) is placed in the path of light as illustrated in Fig. 46.

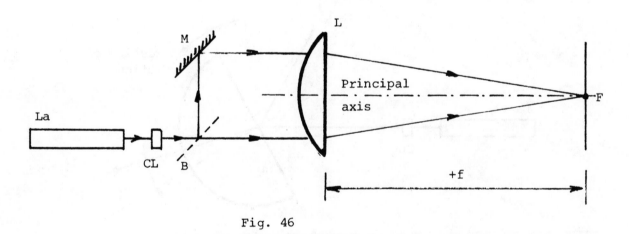

Fig. 46

The point of intersection of the refracted ray streaks is the principal focus (F) of the convex lens. The distance on the principal axis from the lens to F can be measured and is the focal length (f).

(7-B) <u>Focal Length of a Negative (Concave) Lens.</u> The setup is the same as in (7-A). Similarly, two parallel ray streaks are employed. In this case, virtual rays are drawn in as extensions of the refracted rays on the right side of the lens until

they intersect at F on the principal axis. The distance between the
center of the lens and the focal point F is its focal length (f). It
should be noted that the distances designated f have a positive sign
for converging (convex) and a negative sign for diverging (concave)
lenses (see Fig. 47).

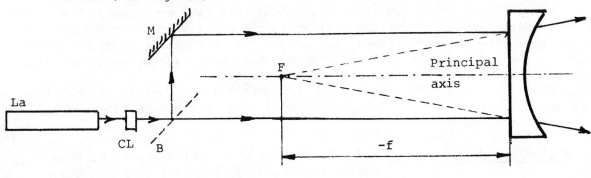

Fig. 47

(7-C) <u>Spherical Aberration.</u> The setup is similar to that used in (7-A). A
plano-convex lens is now positioned with its
principal axis outside the two parallel ray streaks. Also note that
in this exercise the curved surface of the lens model is facing away
from the incoming ray streaks. Ray #1 is a paraxial ray and Ray #2
is a marginal ray. The shift of focus for the two rays is quite
noticeable and they are indicated by the little arrows at F_1 and F_2.
By translating the lens in the direction of the arrows, the shift of
focus for paraxial and marginal rays is clearly demonstrated (see
Fig. 48). Note that when the lens is turned around so that the convex
surface faces the incoming ray streaks, the spherical aberration is
markedly reduced.

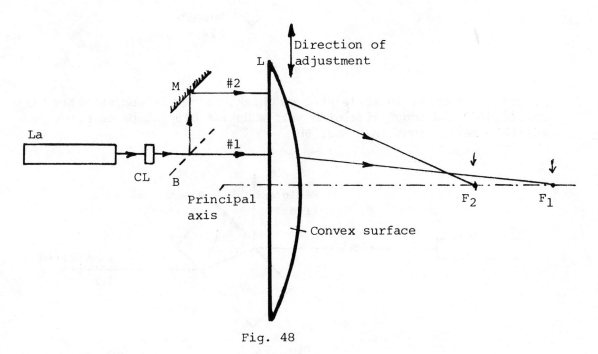

Fig. 48

Beamsplitters divide an input light beam into two components. The output components emerge in different directions - usually at right angles to each other. The relative intensities of the two emergent light beams are fixed by the characteristic of the beamsplitter, that is, a beamsplitter is capable of reflecting and transmitting an incoming beam in a certain characteristic ratio. This reflection/transmission ratio (R/T) is usually expressed in a percentage. A description of some of the more often used beamsplitters follows...

A) <u>Flat Glass Beamsplitters</u> are semi-mirrors which reflect part of the incident light (absorb some light) and transmit the rest. The plate glass should be parallel to 20 seconds and the surface flat to 10 rings. The standard plate glass beamsplitters are coated with a semi-transparent reflecting layer of metal, usually silver or aluminum, to reflect and transmit about equal amounts of the input beam (see Fig. 49).

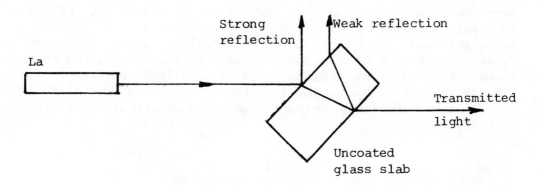

Fig. 49

If the glass plate is at least 6 mm thick, it can be conveniently half-shielded on the input side with self-adhesive black tape to block out multiple reflections (see Fig. 50).

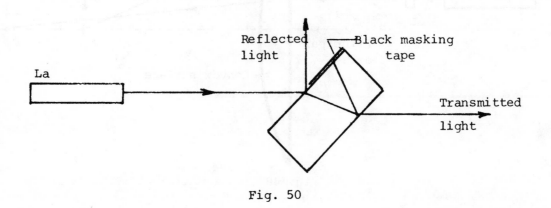

Fig. 50

A microscope slide is too thin to cut out multiple reflections and may give undesirable macroscopic 'wedge' fringes.

Figure 51 illustrates a beamsplitter utilizing a back silvered thick glass block or slab. One beam results from the front surface air-glass interface and the other beam from the silvered back surface reflection. Beam #1 is much less intense than Beam #2.

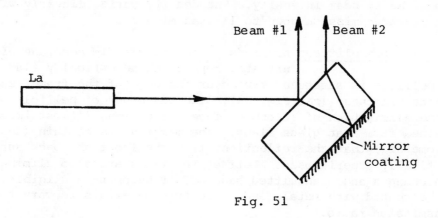

Fig. 51

Fig. 52, below, depicts a variation of Fig. 51.

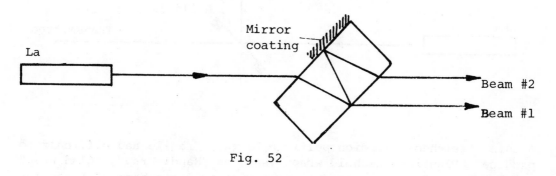

Fig. 52

Only half of the input surface is silvered. Beam #1 is about 90% while Beam #2 is about 3% of the input beam intensity.

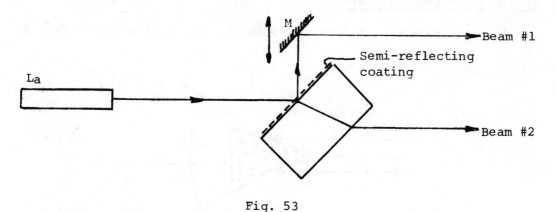

Fig. 53

Figure 53 is a still further variation of the flat glass beamsplitter. The glass surface facing the light source is coated with a semi-reflecting metal layer. The front-surface mirror M is optional. It is added to re-direct the reflected beam and/or as a convenient means to vary the separation of the two split beams. Beams #1 and #2 are about the same intensity.

When a continuously increasing density of inconel is evaporated on a plane parallel optical flat one obtains a variable density beamsplitter. The sensitive coated surface is overcoated for protection. By translating the variable beamsplitter across the laser beam, fine adjustment may be made as to beam intensity. The density varies linearly with length - from 96% transmission down to 1% transmission.

B) <u>Pellicle Beamsplitters</u> are made of tough, elastic membrane of cellulose nitrate, mounted on an optically flat metal frame. The pellicle is stretched taut over the edge of the frame, ensuring surface flatness (see Fig. 54). The advantage of pellicle beamsplitters is the elimination of second surface reflections ('ghost images') as obtained from flat glass slabs. The membrane is so thin (about 7-8 micrometers) that the reflections from the front and back surfaces are practically superimposed. Refractive errors are also eliminated, with path changes and transmitted-beam offset becoming negligible. Pellicles may be coated with metals and dielectric materials to vary the reflection/transmission ratio.

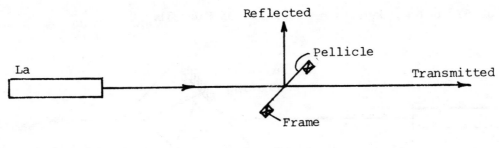

Fig. 54

A thin stretched collodion pellicle is very fragile and difficult to produce. Plastic household wraps (such as "Handi-Wrap", "Glad Wrap" or "Saran Wrap") can be used to improvise pellicle beamsplitters for demonstration exercises.

C) <u>Single Prism Beamsplitters</u>.

A collimated beam of light is incident on a prism as shown in Fig. 55.

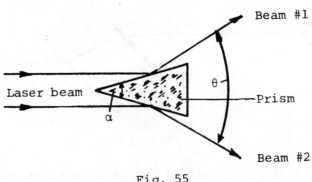

Fig. 55

Part of the input beam is reflected from one face and part from the other face of the prism. The angle θ between the two reflected beams

(Beam #1 and Beam #2) is twice the angle α between the two reflecting surfaces of the prism.

D) <u>The Cube (double-prism) Beamsplitter</u> is formed by bonding together two right angle prism with appropriate coatings on the interface. The unit thus forms a cube (see Fig. 56).

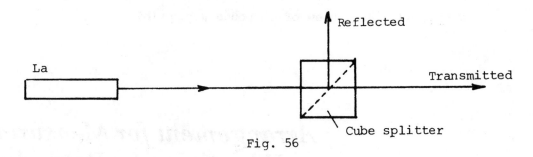

Fig. 56

The hypotenuse of one prism is provided with a thin semi-transparent reflecting layer of metal, so that part of the light is reflected and part transmitted. These cubes usually have reflection and transmission values of 30 to 40% each. The rest of the light is absorbed by the reflectance coating. Refraction aberrations produced as the light enters are compensated by opposite errors on emergence. The fragile semi-transparent film is protected by the prisms and there is no problem with "ghost images."

E)   <u>Beam Multiplier Assembly.</u>

A laser beam can be split into several beams for mirror and lens demonstrations with the help of a compound optical wedge. (Reference: W.H. Porter, The Physics Teacher, p239, Apr. 1974).

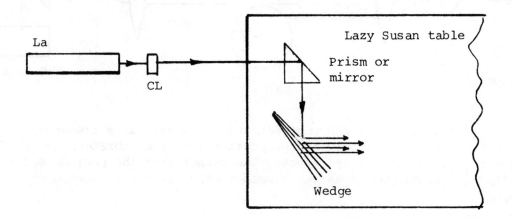

Fig. 57. Top view of setup (a)

Four or five microscope slides are stacked and taped together at one end. The slides are separated at the other end with strips of cardboard and the sandwich is taped together. The reflected components give slightly divergent rays but they are parallel enough for demonstration purposes.

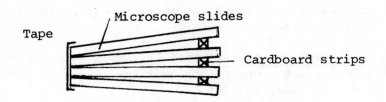

Tape

Microscope slides

Cardboard strips

Fig. 57. Side view of compound wedge (b)

Arrangement for Measuring Vibrations of a Rotor Axle

In Figure 58 the laser beam is expanded by means of the optical system L_1 and L_2. The expanded and collimated laser beam travels toward the test object T (rotor axle, elastic vibrator reed, etc.) and the light rays which pass by the object are focused through L_3 on a photodiode PD or other type photodetector.

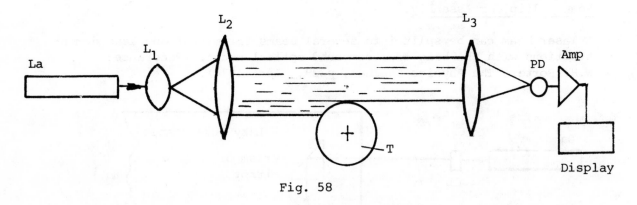

Fig. 58

The light flux reaching the photodetector will vary at a frequency equal to the vibration frequency and proportional to the vibration amplitude of the rotor axle or other test object. The output from the photodetector is amplified and displayed on an oscilloscope where it may be viewed or photographed.

Reference:

A.T. Kornev and G.N. Mironova, Izmeritel'naya Tekhnika, n5 p36 (May '70)

Optical Lever

A small mirror is attached to an object. If a laser beam is reflected from this mirror onto a distant wall or screen, any small motion of the object will be greatly magnified in the motion of a bright red spot formed by the reflected beam. This comprises an optical lever.

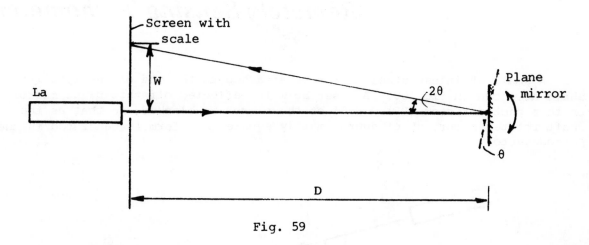

Fig. 59

If the mirror is tilted through an angle θ, this will cause the reflected laser beam to rotate through 2θ. The optical arm is D, and the width of deflection is W. The relationship between these parameters is:

$$W/D = \tan 2\theta.$$

For small angles, the small angle approximation is made: $\tan 2\theta \simeq 2\theta$ and $2\theta \simeq W/D$.

Figure 59 shows a simple technique of measuring the angular rotation θ of a mirror. The principle of the optical lever is used in the light spot galvanometer to demonstrate in a lecture hall small currents from thermopiles or from other sources of very small currents. The galvanometer has a mirror attached to the moving coil to indicate its deflection. Very small changes in current can be detected as measurable light-spot movements. The beam of a 1 mW He-Ne laser forms a bright spot on the wall or screen to be clearly visible even in a well lighted auditorium.

The optical lever can be used to display small changes in length that occur in a thin wire when the force pulling the wire is changed. The technique can be used to calculate the Young's modulus of the wire.

A moving-coil galvanometer and a laser beam can form a scanning system. A repetitive sawtooth shaped signal is fed to the galvanometer, causing the mirror to oscillate. The laser beam reflected from the mirror scans the document. A photodetector receives the light reflected from the document and produces an electrical output signal. Such a laser-based scanning system has, - among other things,- greatly increased the transmitting speed of facsimile systems.

References:

Bartlett, A.A. and Stoller, R., Amer J Phys v41 p1116 (Sep '73)
Kshatriya, A., Phys Teach p501 (Nov '74)
Tenney, H.M. and Purcupile, J.C., Electro-optical Syst Des v7 p40 (Oct '75)
Brosens, P., Opt Engineering v15 p95 (Mch/Apr '76)

Remotely Sensing Tachometer

A small stainless steel mirror is attached to the end of the rotating
shaft as shown in Fig. 60. A laser beam is reflected off this plane mirror
on to a photodetector (PD). Half of the mirror is blacked, so that as the
shaft rotates a current of approximately square wave form is generated in the
photodetector.

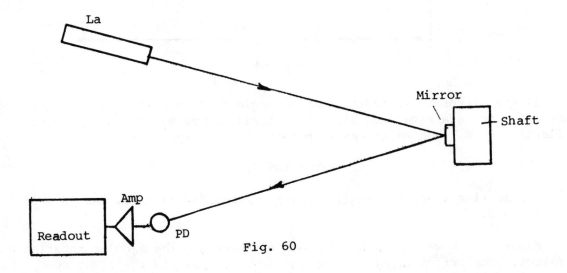

Fig. 60

The photodetector and amplifier produce an alternating voltage signal of
the same frequency as the spin rate of the rotating member. The output of the
amplifier may be fed to either an oscilloscope or to a frequency-to-D.C.
converter to provide a real-time plot of spin rate versus time.

Approximating the Radius of
Curvature of Small Concave Mirrors

For the investigation of reflections from curved surfaces, the He-He
laser is an excellent light source. It can be employed to approximate the
radius of curvature and focal length of concave mirrors whose radius and
focal length are unknown.

The optical setup is shown in Fig. 61. First, the laser is in Pos. #1

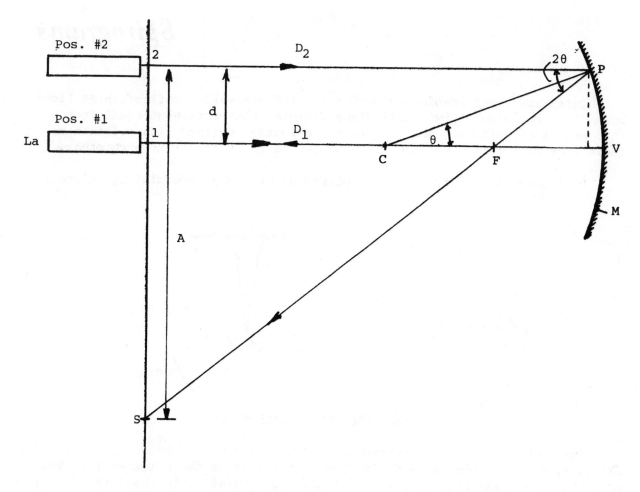

Fig. 61. Measuring the radius of curvature.

and the beam originates at point 1. The beam strikes the mirror at V and the mirror is so adjusted that the beam is reflected back on itself to point 1. Thus, the beam path is 1-V-1. Second, the laser is moved to Pos. #2. This results in the laser beam being translated a transverse distance d from point 1 to point 2. The position of the mirror is held fixed. The laser beam is reflected off the mirror surface at P and the reflected beam strikes the screen at S. The path taken by the beam will be 2-P-S. An angle of 2θ is formed by the translated laser beam and its reflection, θ is the angle formed by the radius of curvature and the translated beam. D_1 is the distance between the screen at 1 and the mirror at V. D_2 is the translated distance from 2 to P. The distance A is measured from 2 to the center of the spot S on the screen.

The radius of curvature is related to d, D and A. We assume that $D_1 \simeq D_2 = D$, and A is much smaller than D. By constraining d to small values, the radius of curvature $R \simeq 2dD/A$. The focal point of a concave mirror is halfway between the center of curvature, C, and the vertex, V, of the mirror. Thus, the focal length is equal to one-half of the radius of curvature ($f = R/2$).

J.D. Evans, Appl Opt v10 p995 (Apr '71)
J.D. Evans, Appl Opt v11 p712 (Mch '72)

This experiment involves a device for the sinusoidal deflection of light beams. It consists of two small plane mirrors which rotate at equal speeds in opposite directions. The mirrors are slightly inclined with relation to their axes of rotation but the axes themselves are parallel to each other.

A circular scan pattern can be generated with one "wobbulating" mirror.

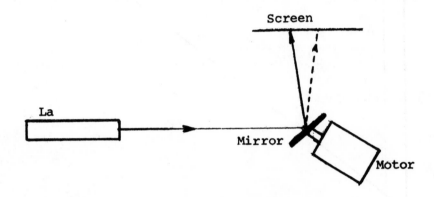

Fig. 62. The "wobbulating" mirror

The wobbulation of the mirror is caused by its non-perpendicular or angularly offset alignment with respect to the axis of the motor shaft. The incident laser beam is inclined at a fixed angle relative to the axis of rotation of the tilted mirror. The angle of incidence, however, between the beam and the normal to the tilted mirror varies in accordance with the instantaneous position of the rotating tilted mirror. This variation results in the reflected beam forming a conical scan. If a viewing screen is positioned perpendicular to the axis of the conical scan, the reflected laser beam will form a circular pattern.

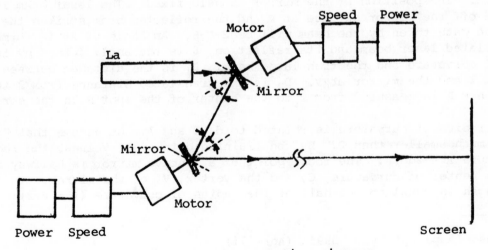

Fig. 63. Counter-rotating mirrors.

Figure 63 is a diagram of the experimental setup. A pair of mirrors are counter-rotated about parallel axes and are tilted equally (α) with respect to the axes so that an incident laser beam is reflected by the two mirrors and emerges with a sinusoidal oscillation. The sinusoidal line scanners are so positioned that the emerging beam from the first scanning mirror is incident upon the second.

If the two mirrors are interconnected in such a manner as to produce synchronous and in-phase rotation, then the circular deflections of a laser beam produced in each mirror are mutually cancelled and the scanning movement of the laser beam is a back and forth one along a straight line. In our experiments an individual drive for each mirror was provided, with provision for adjustment of the drive motors to insure synchronous movement of the mirrors in correct phase relation.

Many other scan patterns can be obtained by (a) varying the mirror tilt angles, (b) the mirror rotation speeds, (c) the mirror rotation phase angles. Some of the Lissajous patterns (Spirograms) obtained with the experimental setup are shown in Fig. 64.

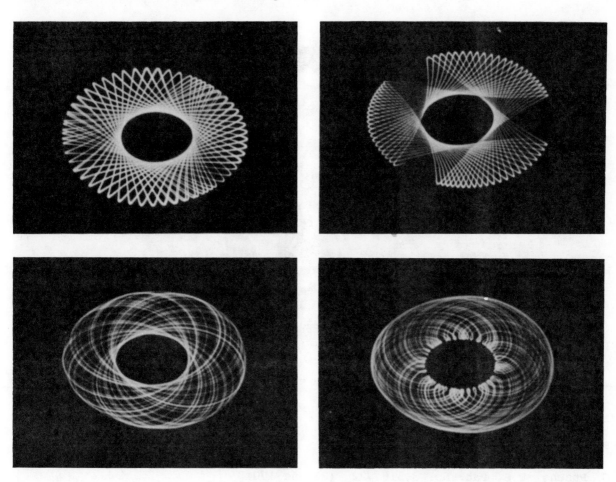

Fig. 64. Spirograms

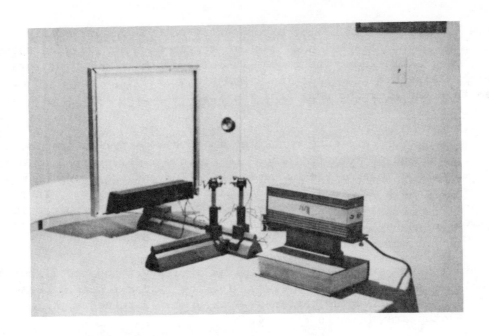

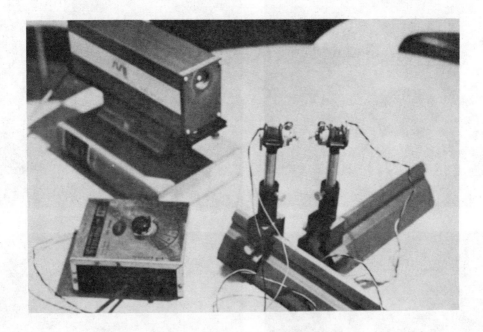

Fig. 65. Spirogram experimental setup

For further reading:

G. Pusch, U.S. Pat. No. 3,516,722 (23 Jne '70)
W.H. Keene and C.M. Sonnenschein, U.S. Pat. No. 3,619,028 (9 Nov '71)
I. Gorog, U.S. Pat. No. 3,630,594 (28 Dec '71)
N.L. Stauffer, U.S. Pat. No. 3,776,639 (4 Dec '73)

Laser Optical Plummet

A collimated He-Ne laser beam gives a natural visible reference line for any type of alignment survey.

For the vertical alignment of high building constructions, deep mining shafts, or well drilling, the laser beam must be set in a vertical direction see Fig. 66).

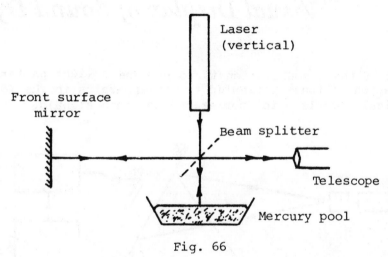

Fig. 66

The laser optical plummet uses the autocollimation technique shown above. The laser beam is projected onto, and reflected from, a mercury pool and directionally adjusted until the outgoing and reflected laser beams coincide. The first surface mirror is adjusted by means of an autocollimating telescope to be perpendicular to the axis of sight of the telescope. Then the various components are adjusted by means of precision gears until the two laser beams (reflections from the mirror and from the mercury), coincide with the center of the reticle of the telescope.

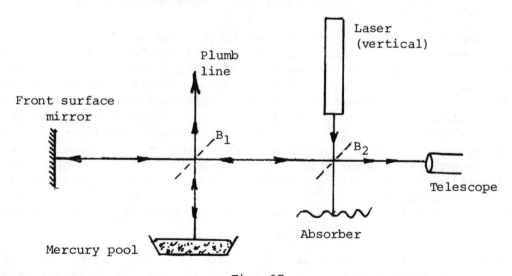

Fig. 67

The plummet can also be used in an upward direction if two beamsplitters are used (see Fig. 67).

_____

For further reading:

A. Chrzanowski, F. Ahmed, and B. Kurz, Appl Opt v11 p319 (Feb '72)

Visual Display of Sound Dynamics

The object of the demonstration is to produce a light pattern for each tone or combination of tones presented to the apparatus in the form of time-varying electrical signals which correspond to sound signals.

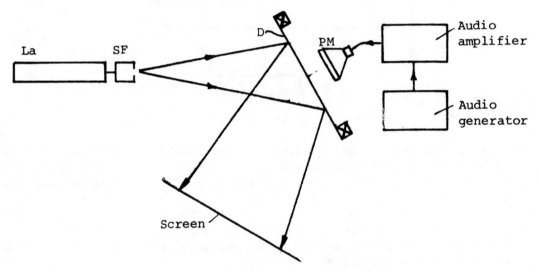

Fig. 68. Experimental setup.

A diaphragm, D, is stretched in a suitable frame and a speaker, PM, is mounted concentrically with the diaphragm. PM is the output speaker of an audio signal generator/amplifier system. The diaphragm may be a thin plastic membrane with a mirror surface, such as vacuum deposited aluminum. Saran-Wrap (Dow Chemical Co.) was one material found satisfactory along with a number of other thin plastic films tried. The filtered (SF) and expanding laser beam is directed toward the mirror surface of the diaphragm from which it reflects to a viewing screen. The diaphragm is made to vibrate in its characteristic modes by sound pressure waves from the loudspeaker. The vibration of the membrane will distort the mirror surface which will produce characteristic light patterns on the screen. The physical mode patterns are determined by the properties of the membrane and the boundary condition of the membrane holder.

Instead of a thin plastic membrane, a thin soap film may be used. A soap solution found satisfactory was one made of equal parts of distilled water, pure glycerine, and Trend (a commercial household preparation). Soap films of different shapes are used to point up the effects of different boundary conditions (see Fig. 69).

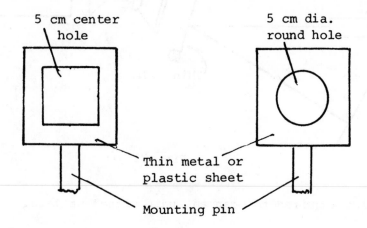

Fig. 69. Soap film frames.

A soap film is prepared by dipping the film's holder into the solution described. After placing it in position, the film should be allowed a few seconds to drain. The light beam then is reflected from the surface of the square soap film to the screen. As the frequency of the driving wave is varied, the positions of the radial nodes are observed to shift in accordance with the requirements of the boundary conditions. With the circular film in place the experiment is repeated. Figure 70 shows a variation of the experimental setup...

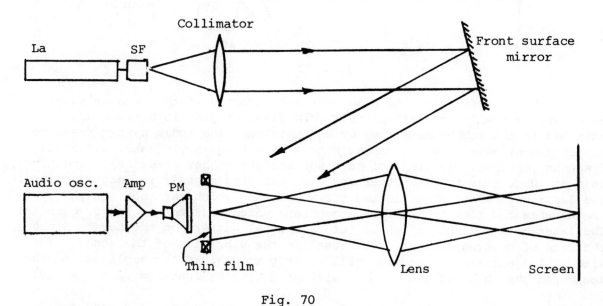

Fig. 70

Figure 71 shows a modified system (see next page).

The very narrow beam of the laser is directed to one point on the thin film diaphragm where the vibrations in the diaphragm will deflect the laser beam into spatial patterns on the screen. In this case the pattern will appear to be a line figure because the retinal retention of the eye is much

51

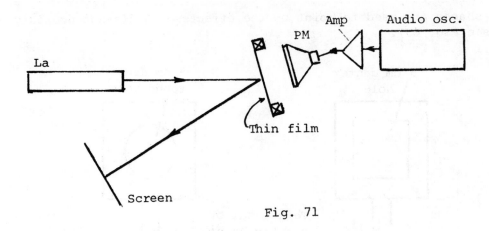

Fig. 71

longer than the time required to scan an audio signal pattern.

An alternate system is shown in Fig. 72.

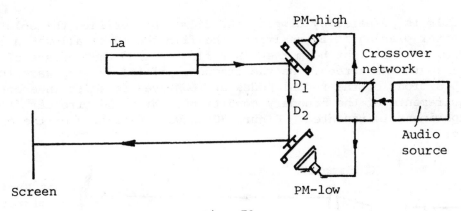

Fig. 72

In this modification the membranes are composed of thin rubber sheets which are optically non-reflective. Thin glass or plastic mirrors are attached to the rubber membranes by an adhesive. The audio system feeds to the crossover network which connects to a high frequency speaker and a low frequency speaker. In front of each speaker are rubber membranes respectively held at 90 degrees to each other and each carrying a small reflective mirror. The laser beam is first reflected off a membrane which responds to high frequencies and then to and from a membrane which responds to low frequencies. The laser beam will be reflected into a spatial pattern on the screen since the rate of membrane vibration is equal to the vibration of the sound. The size of the pattern so produced will be proportional to the amplitude of the sound and the shape of the pattern will be a complex function of the sound.

Figures 73 a) and b) were produced from a point source directed to Saran-Wrap at 3,000 Hz and 300 Hz respectively. Figs 73 c) and d) show line patterns produced by a high and a low frequency sound signal respectively, with the laser beam reflecting off a mirror mounted on a rubber membrane.

(continued on next page)

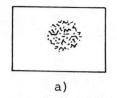 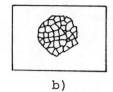 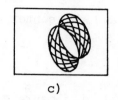 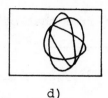

a)   b)   c)   d)

Fig. 73. Typical patterns

Reference:

L.G. Cross, U.S. Pat. No. 3,590,681 (6 July 1971)
T. Campbell, Phys Teach v10 n5 p283 (May '72)

Scanning a Focused Light Beam Over a Wide Flat Field

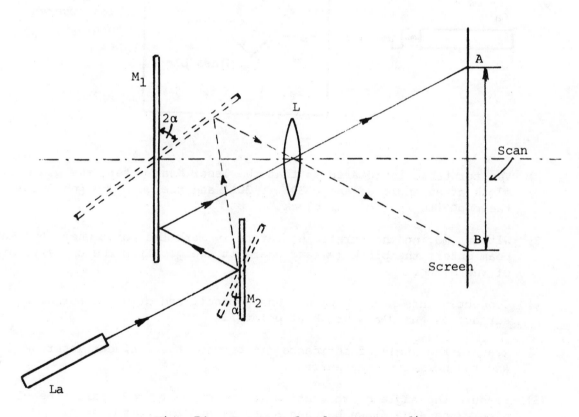

Fig. 74. Scanner for laser recording system.

Two front surface mirrors are positioned to intercept the input laser beam and to scan it over field AB. Lens L is positioned in the beam's path and focuses the beam on a screen or recording medium. Mirrors M_1 and M_2 are oscillating or continuously rotating in a synchronized manner but the speed of rotation of M_1 is double that of M_2. The axis of rotation of M_1 is equidistant from that of M$_2$ and the center of lens L.

Reference:

A.G. Dewey, IBM Technical Disclosure Bulletin v17 n9 p2743 (Feb '75)
R.E. Hopkins and M.J. Buzawa, Opt Engineering v15 p90 (Mch/Apr '76)

Refractive Index Measurements

A)   Refractive index of a rectangular glass block.

1) The outline of the glass block is drawn on a piece of graph paper
which rests on the lazy susan turntable (see Fig. 75).

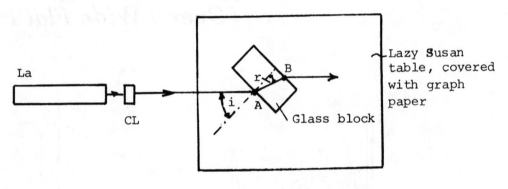

Fig. 75

2) The turntable is rotated so that the laser beam enters the glass
block at an angle of about 30° (i=30°) and the beam emerges from
the opposite side of the block.

3) Without moving the turntable, carefully mark the point where the laser
beam enters the block (A) and where it emerges from the opposite side
of the block.

4) Connect points A and B with a sharp pencil and draw the normal to the
surface of the glass block at point A.

5) Measure the angle of incidence (i) between the incident laser beam
and the normal to the surface.

6) Measure the angle of refraction (r) which is formed inside the glass
block (the line drawn between A and B) and the normal at A.

7) Calculate the index of refraction using Snell's law:

$$n = \sin i \ / \sin r$$

8) Repeat the above procedure after rotating the turntable so that the

angle of incidence is changed for each trial. Note the effect on
the size of the angle of refraction as the angle of incidence is
increased or decreased. For each angle of incidence, calculate the
index of refraction of the glass. Confirm Snell's law which states
that on refraction, the ratio of sin i to sin r is a constant
depending only on the nature of the media.

B)   Refractive index of transparent liquids.

Repeat the previous exercise by replacing the glass block with a hollow
glass cell with flat, parallel walls. The cell contains the specimen
liquid. Note: For a high-precision scheme read: "The Amateur Scientist,"
in Scientific American, v232 n5 p108 (May '75).

C)   Refractive index of a glass prism.

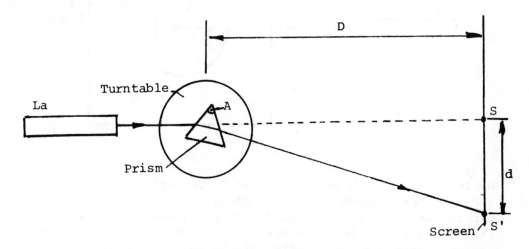

Fig. 76

The light beam from a laser is directed so that it strikes a distant
screen perpendicularly (see Fig. 76). A bright spot appears on the
screen at S.   Next, a glass prism is placed in the path of the laser
beam. The beam deviates and the spot now appears at S' on the screen.
When the turntable on which the prism rests is rotated, the spot S' moves
on the screen. As the spot moves to and fro on the screen, find the
minimum deviation by measuring the least distance (d) between S and S'.
The angle α is found by the formula: $\tan \alpha = d/D$ and the index of
refraction is determined by the formula:

$$n = \frac{\sin (A + \alpha)/2}{\sin (A/2)}$$

where A is the apex angle of the prism and α is the angle of minimum
deviation.

Note: The distance D and d should be quite a few meters and the prism
angle A should be known precisely.- Instead of a glass prism, a prism
shaped glass or plastic vessel can be used to find the refractive index

of various liquid specimens. - For further reading see: N. Naba, The
Physics Teacher, p241 (Apr '73).

D)   Refractive index of liquids measured with a semicircular clear
plastic cell.

The concept of the critical angle is used to measure the index of
refraction of a liquid. The critical angle is that angle of incidence
for which the angle of refraction is 90°.

A semicircular clear plastic cell (available from Macalaster Scientific
Co., Cat. No. 80128) is filled with the liquid specimen.

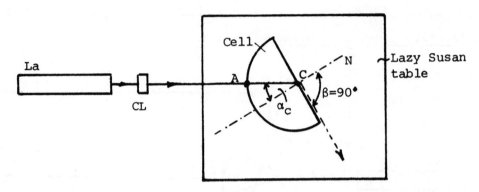

Fig. 77

The incident laser beam is aimed at C (center) of the cell (see Fig. 77).
Since the beam enters the liquid along a radius there is no refraction
when the beam enters the liquid. At C the emerging beam is refracted
with an angle of refraction greater than the angle of incidence. As the
turntable carrying the plastic cell is rotated and the angle of incidence
α is made larger, the angle of refraction eventually becomes 90°, and the
refracted beam travels along the straight surface of the cell. Upon
further rotation of the plastic cell the laser beam is totally reflected
within the plastic cell.

In order to measure the critical angle, first, the outline of the semicircular cell is drawn on the piece of paper which rests on top of the
lazy susan table. Second, the turntable is carefully adjusted for the
critical angle. At the critical angle the refracted beam disappears.
Third, points A and C are marked on the paper. Fourth, the cell and the
paper are removed from the turntable and the line AC is drawn. Fifth,
the angle between AC and N (normal) is the critical angle (α_C) and it is
measured with a protractor. From Snell's law the index of refraction n
at the critical angle α_C is given by $n = 1/\sin \alpha_C$.
For further reading see: E.D. Noll, The Physics Teacher p309 (May '73)

" " " " " " " " " "

"Double-Spot" Technique for Measuring Small Wedge Angles

One of the applications for the optical arrangement shown in Fig. 78 is the measurement of very small wedge angles in optical flats. The technique involves measuring the angular separation between the beams reflected from the front and back surfaces of the flat.

A low-power He-Ne laser is used as a source of light. To demonstrate the technique, one half of a biprism is used as a wedge. When the direct laser beam impinges upon the wedge, two spots appear on the screen. From the spot-separation distances, d, we can determine the wedge angle, α.

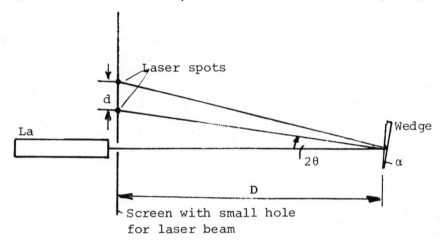

Fig. 78

The relation between d and α is easily obtained. Small angle approximations can be used in the calculations and it is sufficient to use an average value for the wedge thickness, t,

$$\alpha = (d/2 - \theta t/n)/(t + nD) \ldots\ldots\ldots\ldots (1)$$

where θ is the angle of the incident laser beam to the first reflecting surface and n the index of refraction of the material.

In using the double-spot technique, it is useful to adjust the beam approach angle (θ) to near zero. This is accomplished by so aligning the wedge that the incident laser beam falls almost normal to the front reflecting surface. Thus, the spots are reflected back along a line close to the input laser beam. If the wedge to screen distance, D, is made quite large compared with the average wedge thickness, t, equation (1) simplifies to

$$\alpha = d/2nD \ldots\ldots\ldots\ldots\ldots\ldots (2)$$

In estimating α using Eq.(2) it is good practice to make two measurements of the spot separation, d, the second with the wedge rotated 180 degrees in its plane relative to the first. This eliminates the effect of a non-zero θ, which occurs because one cannot reflect the beam back along its original path and still see both spots on the screen. However, in case of the biprism demonstration, one must make sure to use the same half of the biprism for both measurements!

References:

P. Gallagher and A.J. Rees, Appl Opt v10 p1967 (Aug '71)
J.H. Wasilik, T.V. Blomquist and C.S. Willett, Appl Opt v10 p2110 (Sep '71)

Curving and Bouncing Laser Beam

A light beam is deflected by a prism - that is, by a transparent material of uniform refractive index and varying width. A transparent slab of material (glass or plastic) also deflects a light beam, the amount of deflection being a function of the width of the slab. In case of a mirage which forms over hot sand, the light beam is deflected by the refractive index gradient in the air due to the temperature gradient close to the ground. In such cases as the Fata Morgana, or light reflection on a hot road surface, the light rays bend upward.

The bending of a light beam by refraction with uniform change of the refractive index can be demonstrated with a simple laboratory setup (see Fig. 79). The effect is illustrated by deflecting a laser beam in the mixing zone of two liquids.

The phenomenon is observed in a transparent vessel (fish tank) of about 30-40 cm length. The tank is filled to a depth of about 3-4 cm with the kind of concentrated rapid fix solution commonly used in photography. This is the denser liquid. A lighter fluid, such as tap water is allowed to form another layer. A thin strip of wood, paper, or cork is floated on the concentrated rapid fix and tap water is added by slowly pouring it down a stirring rod on to the top of the floating cork to minimize initial mixing of the two liquids. The two suggested liquids mix appreciably within two hours.

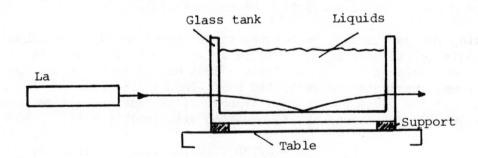

Fig. 79. The curving laser beam.

The laser beam enters the glass tank about 1-2 cm above the tank bottom. The bending light path develops as the concentrated rapid fix diffuses into the tap water. A concentration gradient evolves with the highest rapid fix concentration at the bottom of the tank. The rapid fix concentration gradient establishes a vertical gradient of index of refraction.

The bending laser beam reaches the bottom of the glass tank and "bounces" by total reflection at the glass-air interface at the outer surface of the glass tank. The totally reflected beam bounces upward first, then begins to curve downward again provided the glass tank is long enough to observe the effect.

If the laser beam is expanded with the help of a beam-expander/collimator, the phenomenon of periodic focusing of the laser beam can be observed. With a 5 mm diameter beam the effect can be seen fairly easily. The periodic focusing occurs only in the vertical plane; the width of the beam in the horizontal plane remains unchanged.

References:

A. Sommerfeld, "Optics, Lectures on Theoretical Physics," (Academic Press, New York, 1964) vol. IV, pp338-352
W.M. Strouse, Am J Phys v40 p913 (Jne '72)

Recording and Measuring Refractive Index Gradients

The spatial variation of the refractive index in the mixing zone of two liquids can be determined from the observed and recorded deflection of a laser beam.

The beam from a low-power He-Ne laser is passed through a glass or plastic rod or 3-5 mm diameter. The rod is mounted at 45-degree angle to the vertical. The light emerging from the rod is fan shaped. This thin blade of light then passes through a small glass cell (approx. 5x5x1.5 cm) containing the liquids. When the cell is empty, the fan shaped light rays pass through the cell undeflected and give a straight line illumination at 45-degree to the vertical on the viewing screen (see Fig. 80).

First, the denser liquid is poured into the cell. For demonstration purposes, this liquid may be concentrated photographic rapid fixer. The cell is filled halfway. A small piece of paper, wood or cork is floated on this liquid surface and the lighter liquid (tap water) is slowly guided onto the top of the cork by a mixing rod. This slow pouring method will minimize initial mixing of the liquids.

A light ray passing through a medium whose refractive index varies with depth will be bent into an arc. In the experimental arrangement the fan-shaped

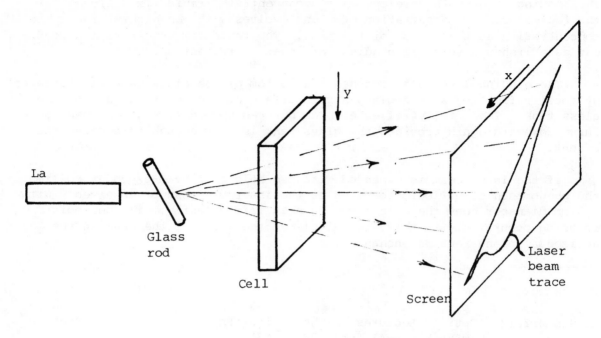

Fig. 80. The experimental setup.

rays of the laser beam pass through the liquids at different depths, "y" and are deflected through different diatnces, "z", according to the local value of dN/dy. The deflected beam intersects the screen in a curve of the shape shown in Fig. 81.

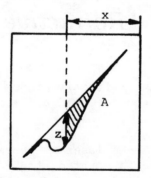

Fig. 81. Shape of the deflected
beam.

The refractive index N(y) can be found from this curve. The method is described in the reference given below.

_____

Reference:

A.J. Bernard and B. Ahlborn, Am J Phys v43 n7 p573 (Jly '75)

"""""""""""""

Direct Shadow Demonstration
(Shadowgraphy)

This demonstration illustrates a simple ray optic method by which very small changes in the refractive index of air can be made visible. This method of investigating the flow of gases is widely used in the study of ballistics and in aeronautical and combustion engineering.

The arrangement for the visual observation of the direct shadow technique is shown in Fig. 82.

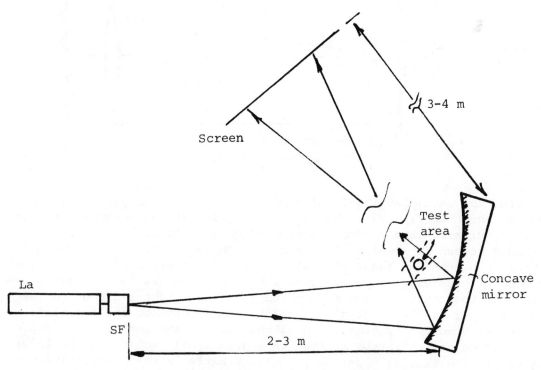

Fig. 82. Direct shadow demonstration (not drawn to scale).

The laser beam is filtered (SF) and the divergent beam is incident on a front-surface concave spherical mirror (for example, Edmund N. 40913, 25" focal length). Mirrors suitable for telescopes are satisfactory. The mirror is slightly tilted so that the reflected light falls on a distant screen. Changes in the density and refractive index in the test area result in corresponding changes in illumination on the screen (the differences in density of the image are proportional to the derivative of the gradient in refractive index).

Since the light source and the screen cannot both be normal to the mirror, some coma is introduced by the skewness of the components. The test field must be placed well out in the converging beam from the mirror.

Some ordinarily invisible things that can be seen with this simple

method are: 1) the hot air above a match or miniature candle flame, 2) the
air currents rising from a hot electric soldering iron.

Determination of Focal Length of a Convex Lens by Autocollimation

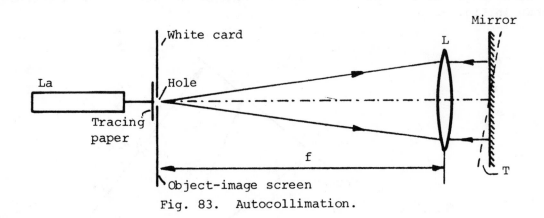

Fig. 83. Autocollimation.

To find the focal length, f, of a positive lens, the apparatus is
arranged as shown in Fig. 83. The laser, the object-image screen, the lens,
and a front-surface plane mirror are mounted on an optical bench. The
object-image screen is a white, 7.5 cm x 12.5 cm (3" x 5") filing card. A
small hole is punched in the center of the card and a piece of tracing paper
is fastened with adhesive tape to cover the hole. The raw laser beam is aimed
to impinge on the center of the tracing paper. The bright red spot on the
translucent paper is the "object". The lens to be measured and the plane
mirror are set up close together. The light leaving the lens is interrupted
by the mirror whose reflective surface is approximately square to the laser
beam axis. If the tracing paper is situated exactly in the principal focal
plane of the lens and the mirror is exactly at right angles to the laser beam
axis than the light rays are reflected back exactly upon their original paths.
If, however, the mirror is inclined at a small angle, T, to the axis, the
incident light rays are reflected back at an angle, 2T, and the image of the
bright red spot will be formed on the white card screen instead of the tracing
paper. By sliding the lens-mirror combination back and forth along the axis
a sharp image of the bright spot is obtained on the white card. The distance
between the card (object-image screen) and the lens is the focal length.

" " " " " " " " " "

Laser Beam Generates
Conic Sections

Conography is a new space-age technology which uses conic curves to digitize and store graphic and symbolic information. The sections of a right circular cone (cone sections) are: a) circle, b) ellipse, c) parabola, d) hyperbola. The following exercise shows a method of generating conic sections (or simply "conics") with a laser beam (see Fig. 84).

Place a white cardboard screen near the end of a thin (2-4 mm) and round transparent rod (glass, lucite or plexiglas) about 30 cm or so in length. The laser beam is directed obliquely at the rod. Since the nominal width of the raw laser beam is less than the width of the transparent rod, a 15-20 cm focal length positive lens, L, is used to expand the laser beam. At the intersection of the laser beam and the rod, the beam width should be slightly larger than the diameter of the rod. After reflection, the locus of the light rays forms a hollow cone which will appear as a narrow ring on the screen held at right angles to the axis of the rod. The cone axis coincides with the axis of the transparent rod. The half angle of the cone will be equal to the angle of incidence of the laser beam impinging on the rod.

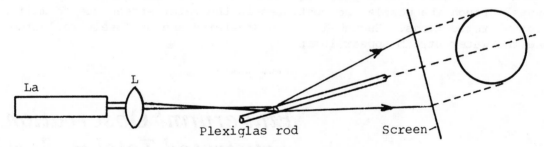

La   L   Plexiglas rod   Screen

Fig. 84. Generating conic sections.

At the beginning of the exercise, when the screen and the rod are right-angled to each other, a circle is formed on the screen. As the screen is tilted relative to the rod the circle changes shape and it becomes an oval-like curve called an ellipse. With still further inclination of the screen - at a point when the side of the cone and the screen become parallel, a parabola is formed. If the screen is tilted further yet, the sides of the parabola will be seen to branch away, and when the screen and the rod are parallel the observed shape will be one half of a symmetrical curve - a hyperbola.

Additional suggested readings:

Selby, P.H., "Geometry and Trigonometry for Calculus," John Wiley and Sons, New York, NY 1975.
Protter, M.H. and Morrey, C.B., "College Calculus With Analytic Geometry," Addison-Wesley Publishing Co., Reading, Massachusetts, 1970.

The Luminous Fountain

In order for total reflection to take place the light at the interface must originate in the medium of greater optical density. In the case of air and water total reflection can take place only when the source of light is inside the water. Figure 85 shows the technique to make the water stream appear luminous.

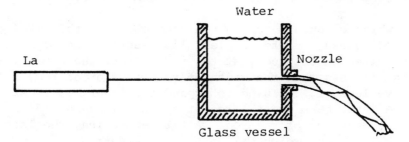

Fig. 85. The luminous fountain.

The laser beam is directed through the water-filled glass (or plastic) vessel at the nozzle from which the water stream is emerging. The laser beam passes through the nozzle and continues in the water stream due to multiple internal reflections. Enough light is scattered and diffusely reflected to make the water stream appear luminous.

" " " " " " " " " "

Fingerprint Observation by Frustrated Total Reflection

When total internal reflection occurs in a medium, radiation can penetrate slightly beyond the reflecting surface into a rarer medium. This phe-- nomenon of penetration can be utilized to frustrate the total reflection completely at the points of intimate contact of an external body brought into close proximity and coupled with the totally reflecting surface from the side of the rarer medium. At the point of contact the radiation will penetrate into the contacting opaque body where it will be absorbed. In this manner total reflection is frustrated at the points of contact.

Fingerprint patterns can be obtained by placing a finger in contact with a prism and viewing the finger through the prism at angles exceeding the critical angle. The surface relief pattern can be seen due to frustrated total reflection (FTR). Because of the presence of the evanescent wave (penetration of the light wave beyond the reflecting surface) total reflection is frustrated (destroyed) at areas of contact with the prism (ridges of the skin) and remains total where contact is not made (valleys and pores). Thus, the ridges are observed as a dark pattern against a light background. The shortcoming of this conventional method of inkless fingerprint observation and photographic recording is the lack of contrast due to bright background illumination and the low intensity of information (the ridges). High contrast fingerprint patterns, however, can be produced with the setup shown in Fig. 86.

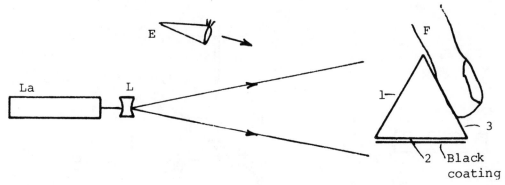

Fig. 86. Fingerprint observation by FTR.

A 60-degree equilateral prism is used instead of the conventional 90-degree prism. The expanded laser beam enters the prism through face 1 and is totally internally reflected at face 3 so as to impinge on face 2 and be absorbed by a non-reflecting homogeneous black paint. This arrangement provides high contrast dark background for fingerprint patterns. The observer at E will normally see only face 2 which is black except where total reflection is destroyed on face 3 by ridges of finger F. Only the ridges on the fingertip will appear brightly illuminated and visible through face 1. The spaces between ridges and the background will appear black as face 2. Application of a very small amount of lanolin or glycerine will enhance the coupling action between the fingertip and surface 3 of the prism.

" " " " " " " " " "

Demonstrating the Penetration of Laser Light Beyond a Totally Reflecting Interface

The penetration of light beyond a totally reflecting interface can be demonstrated as follows: A prism and lens assembly is set to form a variable thickness airgap. The plane surface of a plano-convex lens (borrowed from the Newton's rings apparatus) is optically coupled to the hypotenuse of a 90-degree prism (P_2) using silicone grease (such as Dow Corning 20-057 Optical Coupling Compound). The convex surface is held lightly against the hypotenuse of the other 90-degree prism (P_1). A gap is formed between the prisms which varies gradually from zero thickness in the center to many times the wavelength of light. Figure 87 shows the arrangement.

Two screens are used for qualitative demonstration purposes. They intercept the reflected and transmitted beams. The prism-lens-prism ensemble is slowly moved parallel to the gap direction so that the Newton's ring pattern is scanned by the laser beam. Lens (L) is used to focus the laser beam at the interface. The gradual extinction of the reflected beam and the simultaneous rise in the transmitted intensity is easily observed as the interference pattern is scanned. Quantitative measurements can be made by replacing the screens with photo-detectors.

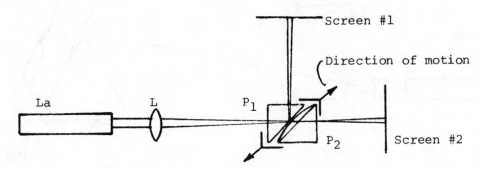

Fig. 87. Penetration of laser light beyond a totally reflecting interface.

The penetration phenomenon is similar to quantum mechanical "tunneling." The device has a practical application as a variable intensity beamsplitter. When a micrometer screw is used to fine-adjust the position of the device relative to the input laser beam, the intensity ratio of the transmitted and reflected beams can be controlled over many orders of magnitude.

References:

Coon, O.O., Am J Phys v34 p240 (1966)
McDonald, W.J., Udey, S.N., Hickson, P., Am J Phys v39 p74 (1971)
   "   "   "   Am J Phys v39 p1141 (1971)

"""""""""

The Plane Parallel Plate as an Optical Micrometer

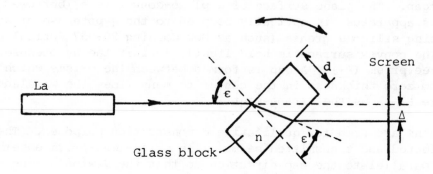

Fig. 88. Optical micrometer.

A laser beam incident at right angles on a rectangular block of glass (or lucite) passes through without displacement. If the block is rotated by an angle ε, the emerging laser beam will undergo a parallel displacement to the incident beam (see Fig. 88).

$$\Delta = d \frac{\sin (\varepsilon - \varepsilon')}{\cos \varepsilon'}$$

The plane parallel glass plate is a simple optical tool for the fine adjustment of a laser beam. Large angular displacements are required to achieve small lateral motions of the beam. The relationship between displacement Δ and rotation angle ε is slightly nonlinear. If adjustments in two orthogonal directions are required, the device consists of two plane parallel glass plates mounted with their rotation axes orthogonal to one another and to the beam.

For further reading:

Goitein, M., Medical Physics v2 n4 p219 (Jly/Aug '75)

The Double Wedge Beamshifter

Two identical glass wedges are combined in this device (see Fig. 89). In Position #1 the two wedges are in contact and form the equivalent of a plane parallel plate of glass. Thus, the laser beam passes through without displacement and appears on the screen at A.

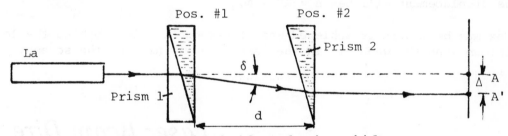

Fig. 89. The double-wedge beamshifter.

If prism 2 is moved along the optic axis by the distance d to Pos. #2, the laser beam will be shifted and appear on the screen at A'. The approximate displacement is given by $\Delta = AA' = \delta d$.

"" "" "" "" "" "" ""

The Mascelyne Type Beamshifter

The collimated laser beam is focused with the help of a long focal length lens (L) on the screen at F (see Fig. 90). A glass wedge is introduced between the lens L and the viewing screen. In Position #1 the laser beam will suffer a deflection δ and will be focused at B. For small wedge angles the displacement will be $\Delta = BF = \delta d_1$.

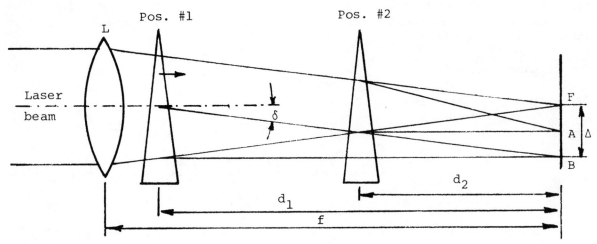

Fig. 90. The Mascelyne-type beam shifter.

In an arbitrarily chosen Position #2 the beam will come to focus at A and the displacement will be $\Delta = AF = \delta d_2$.

Maximum beamshift is achieved when the wedge is right behind the lens and minimum when the wedge is in the immediate vicinity of the screen.

" " " " " " " " " " " "

Laser Beam Director

This device allows a laser beam to be precisely aligned for coincidence in angle and position with a desired line (see Fig. 91).

Two positive lenses (L_1 and L_2) are separated by the sum of their focal lengths ($f_1 + f_2$). They are laterally displaced from the incoming laser beam by the distances x_1 and x_2. The emerging beam is collimated.

When L_1 and L_2 are translated together laterally, the position of emergence ($\bar{X}$) is controlled in two dimensions. The figure shows only displacements in the plane of the paper. When lens L_2 is translated in the lateral plane the angle of emergence (α) is controlled in two directions. The beam diameter may be magnified by selecting the ratio $M = f_2/f_1$.

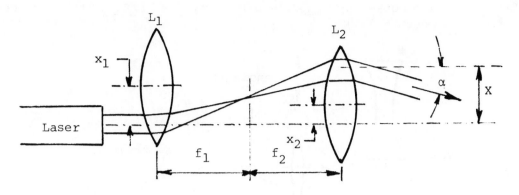

Fig. 91. Laser beam director.

The position of the beam: $X = x_1 \dfrac{(f_1 + f_2)}{f_1}$

The angle of emergence: $\alpha = (x_2 - x_1)/f_2$

Reference:

Harrison, R.W., IBM Technical Disclosure Bulletin v14 n12 p3815 (May '72)

"""""""""""""

"Sliding Lens" Type Rangefinder

Two equally weak lenses, one positive and one negative, form a zero-power lens combination which functions as a variable wedge through the lateral translation of one lens with respect to the other.

A concave and a convex lens of focal lengths $-f$ cm and $+f$ cm are combined so that their optic axes are parallel and displaced from each other by a distance of X cm. A laser beam will pass through such a system without any change of vergence and will be deflected away from the axis of the concave lens by an angle of X/f radians. By arranging for the lateral translation to be variable, an optical unit is formed in which the effective angle of the wedge is variable. Figure 92 shows a rangefinder which incorporates such a unit.

The laser beam is divided by the cube beam splitter. Beam #1 and Beam #2 are made parallel with a front surface mirror or with a 90-degree prism. The length of the base of the rangefinder is b, and D is the distance to be measured; f is the focal length of the lenses forming the variable wedge unit. The unit is set for infinity when X = 0. The minimum range (distance) that can be measured depends on the radius of the lenses. The maximum relative displacement (X_{max}) of the two of the lenses will be equal to twice the radius, less the amount of overlap required. Thus, the maximum angle of deflection will be given by $X_{max}/f = b/D$, or, $D = bf/X_{max}$ (this is the minimum distance measurable.)

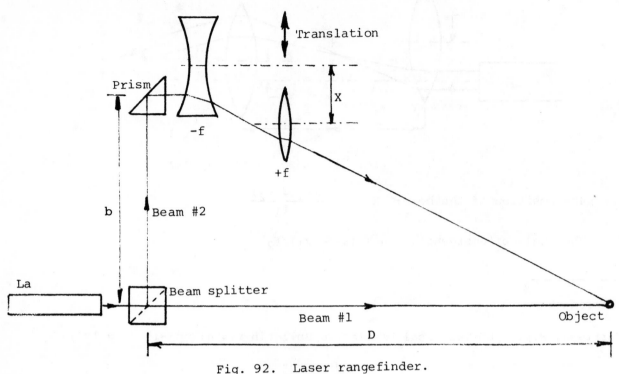

Fig. 92. Laser rangefinder.

Example: For a laboratory (indoor) exercise the minimum range required will be one meter, X_{max} may be 3 cm and b (base) would be about 10 cm.

With these values $f = X_{max}D/b = (3 \times 100)/10 = 30$ cm.

To find the variable displacement X from the infinity position to the distance setting for, say, 10 meters:

$$X = bf/D = (10 \times 30)/1000 = 3 \text{ mm},$$

and for the minimum range of 1 meter the displacement will be

$$X_{min} = (10 \times 30)/100 = 3 \text{ cm}.$$

A graduated range scale can be coupled directly to the positive (sliding) lens. When beams #1 and #2 are brought to coincidence on the object, the distance can be read on the calibrated scale.

_____

For further reading:

Asher, H., J. Sci Instrum v29 p402 (Dec '52)

" " " " " " " " " "

70

Piezoceramic Bender Elements that Control Laser Beams

Piezoceramic bender elements consist of two thin rectangular piezoceramics with their poled axes perpendicular to their electrode surfaces. These piezoceramics are assembled and bonded with epoxy cement to form an alternating polarity bilaminar element. The two layers are polarized in opposite directions. The application of an electric field across the two outer electrodes causes one layer to expand while the other contracts. Thus, the composite element curls and the resulting bending displacement is much larger than the length deformation of either of the two layers. The operation of the bender in response to an electric field is analogous to a temperature change on a bimetallic thermostat.

The curling motion of the bender can control the passage of light and thus act as a light shutter. It can also be utilized as an optical lever to deflect the laser beam (see Fig. 93).

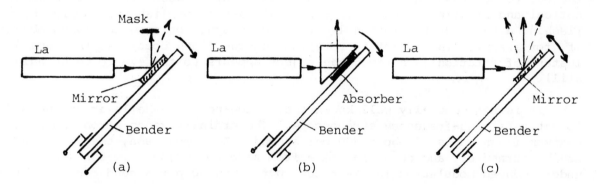

Fig. 93. Piezoceramic laser beam control.

Fig. 93-a shows a simple light shutter. Initially the laser beam is blocked by the mask. When D.C. voltage of appropriate polarity is applied to the bender element which carries a mirror on its end, the laser beam is deflected away from the mask blocking it.

Fig. 93-b shows an arrangement which is based on frustrated total internal reflection. The bender carries a light absorber in optical contact with the prism's reflecting surface. In this state the total internal reflection is destroyed. When D.C. voltage of appropriate polarity is applied to the bender, the absorber moves away from the prism surface and reflection is restored.

Fig. 93-c shows a piezoceramic crystal beam deflector. In this arrangement the laser beam is incident on a mirror mounted on the end of the bender. The beam reflected from this mirror is projected to a distant calibrated screen where it forms a bright red spot. A change in the voltage applied to the bender causes the light spot to move on the screen. Thus, the principle of the optical lever is used in conjunction with the ceramic bender to construct a light-spot voltmeter. When the device is operated with A.C. voltage, it can be used as a low-cost laboratory tool for small beam deflections.

_____

References:

F.W. Kantor, Electronics, p30 (19 Jly '63)
J.M. Fleischer, Appl Opt v10 n8 p1964 (Aug '71)
A.A. Bartlett and R. Stoller, Am J Phys v41 p1116 (Sep '73)
Gulton Industries, Inc., "Piezoceramic Bender Elements" (5M12/74)
J.F. Stephany and I.P. Gates, Appl Opt v15 p307 (1976)

" " " " " " " " " " "

Laser Stroboscope

The function of a stroboscope is to make moving things appear to the eye to be standing still. Strobe action may be accomplished by mechanical, optical and lighting means. Electronic strobes accomplish stop-action by flashing a very short duration light pulse in synchronism with moving objects. If, for example, the object is rotating and the flash occurs at the same instant of rotation each time, the object appears to the eye to be standing still.

It is not generally well known that a low-power He-Ne laser can be made to act as a high-frequency stroboscope. The ordinary xenon stroboscopes have maximum flash rates of about 300 per second. By comparison, the laser beam can be turned off and on at any rate from zero to 100,000 times a second and under reduced modulation levels rates up to 500,000 per second can be achieved.

The modulation circuitry which turns the laser beam on and off is built into the modulatable laser.

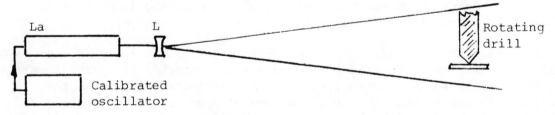

Fig. 94. Laser stroboscope.

For precise high speed strobing, the calibrated audio oscillator is connected across the modulation connector terminals. Best results are obtained using a square wave with a 5-volt output. The modulation input level is made compatible with DTL/TTL digital devices, or can be used in the linear mode over the same interface voltage range by biasing the input to the midpoint of the linear range.

Some applications of lasers with on-off modulation capability:

* Gears and machine parts viewed in slow motion * Vibration analysis
(vibrating reed; high-fidelity tweeter) * To "stop" a flying bullet (high
speed photography) * To "stop" the motion of a high-speed dentist's drill
* To analyze motion in metal cutting and drilling * With the help of a
crayon mark on the shaft motor speeds can be measured (tachometry) * Visual
signaling and warning * Morse-code communications over the laser beam
* Press run count * Reciprocating motion analysis * Camera shutter speed
* Educational applications * Fans * Flutter * Fuel spray * Gravity
demonstration * Loundspeaker studies * Medical applications * Photo-
elastic studies * Projectile velocity * Ripple tank studies * Textile
applications * Physics demonstrations * Data communications * Writing
on photosensitive surfaces.

Two of the simplest objects for stroboscopic demonstrations are a "top"
and a spinning coin.

" " " " " " " " " "

Voice and TV Picture
Transmission Over a Laser Beam

Communication systems are among the most important applications of the
laser. Picture, voice and data transmisssion systems employing a laser beam
can be easily demonstrated in the laboratory or classroom.

A) Voice Communicator System.

Figure 95 shows a typical setup for voice communications

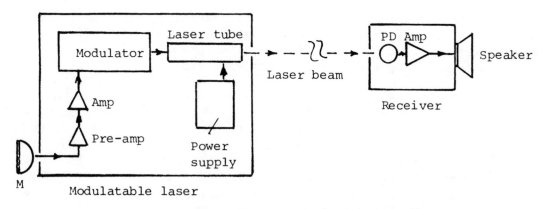

Fig. 95. Voice communicator.

The heart of the system is a low-power modulatable He-Ne laser (such as
Metrologic ML-669) which contains a pre-amplifier, an amplifier, and a current
modulator. The light output from the laser tube is controlled linearly by the
tube current admitted through the modulator circuitry.

The signal source can be a microphone, a signal generator, a tape recorder, or any other audio driver. The only restriction is that the signal input level be less than 1 volt peak-to-peak.

The receiver consists of a photocell, an amplifier, and a loudspeaker. The Metrologic Communicator Kit No. 60-247 comes complete with a crystal microphone which plugs into the back of the ML-669 laser.

The voice signal can be transmitted up to several hundred feet. The working range can be increased considerably by using a 20X collimator at the transmitting end and a converging lens at the receiving end. For an auditorium demonstration of laser voice communications the laser beam could be directed into the Mtrologic 60-255 Photodetector and then the photodetector output used to drive a public address system.

B) Laser TV System.

The system allows sending and receiving black and white video images up to 150 feet over a laser beam. Fig. 96 shows the closed circuit TV hook-up.

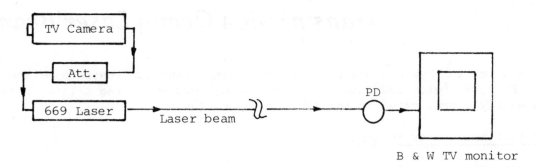

Fig. 96. Laser TV system.

The resolution obtained from the 500 kHz bandwidth of the Metrologic ML-669 modulatable laser is not up to broadcast quality but aside from demonstration purposes it can be used in any black and white closed circuit applications that do not require the maximum resolution.

As shown in Fig. 96, the output from the TV camera is reduced with a standard variable attenuator to the recommended 1 volt peak-to-peak value. The attenuated video signal is fed in to the ML-669 laser. The beam of the ML-669 is picked up by a high-frequency photodetector, PD, (such as the Metrologic 60-255). Since the active portion of the photodetector is very small, only a few millimeters in diameter, it is helpful to use a VTVM or an oscilloscope to monitor the output of the 60-255 Photodetector while aiming the modulated laser beam into the photodetector. Optimum alignment is achieved when the 60-255 output is maximum. The 60-255 can then be used to drive a standard B&W TV "monitor" directly, without any further adjustments.

Figure 97 shows a variation of the laser video link.

The transmitting part is the same as in Fig. 96. At the receiving end the laser beam enters the photodetector through its lens (a phototransistor is used) and the signal is first amplified, then mixed with the output of a

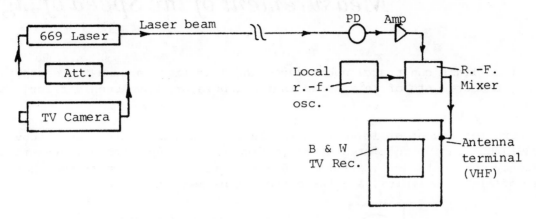

Fig. 97. A variation of the laser video link.

local r.-f. oscillator which is tunable over a 60- to 72 MHz range to permit the laser video link to operate on TV Channel 3 or 4, whichever is not in use in the particular locality. The output of the r.-f. modulator is an amplitude-modulated signal. This signal is fed into an ordinary TV receiver (not a TV monitor!) through its vhf antenna terminals by means of a 300-ohm twin-lead antenna cable.

Devices applying a modulated r.-f. carrier signal directly to a TV receiver's antenna terminals (for example electronic games) are called "Class-1" devices. Since the laser detector/r.-f. modulator falls into this category, FCC requirements must be observed. The two most important regulations are (1) that it is illegal to have the detector's output cable and the TV antenna hooked up to the TV receiver at the same time, and (2) that the device must be formally approved by the FCC. In case of a kit-form Class-1 device, only the manufacturer of the kit is required to obtain approval. No specific restrictions are placed on the laser transmitter. (The Metrologic PE 301 detector/modulator and the PE 500 complete system are approved by the FCC).

The above described laser system works up to a distance of 50 ft. For extended range, a collimator must be used on the laser. Assuming that the raw laser beam has a 1 mRad (milliradian) divergence, the beam diameter will be about 1 meter when measured at 1,000 meters from the laser. Thus, for long distance transmission a collimator (a telescope used backwards) is necessary to keep the beam as narrow as possible. At the detector end of the link a large light gathering lens is used. Lightweight, plastic Fresnel lenses are inexpensive and are well suited for the purpose. They are attached to the detector and focused on the light sensitive surface of the phototransistor. Both the modulated laser and the detector must be rigidly mounted and means for fine adjustment must be provided for precise aiming of the laser.

References:

C.H. Knowles, Popular Electronics p27 (May '70)
G. Punis, and J. O'Donnell, Popular Electronics p32 (Nov '74)

Measurement of the Speed of Light

Since light can travel about seven and half times around the earth in one second, measurements of this speed in the laboratory have been difficult and expensive.

With the advent of low-cost Ne-He lasers (such as Metrologic ML-669 or ML-939) that can be modulated at radio frequencies, and with a low-cost, high frequency photodetector (Metrologic 60-255), it can be demonstrated that the speed of light is finite, and therefore can be measured.

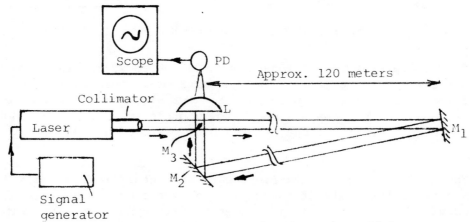

Fig. 98. Measuring the speed of light.

An r.-f. signal generator is connected to a modulatable laser to vary the amplitude of the beam at frequencies up to 750 kHz. A collimator is used to produce a parallel beam about 2 cm in diameter. The beam travels to a distant mirror M_1 and is reflected back to mirror M_2 which directs the beam into the photo-detector. Note in the diagram that there is a tiny mirror M_3 in the beam path which diverts a small portion of the outgoing beam into the photo-detector to serve as a reference beam. Any small mirror may be used for M_3 but the converging lens L must be about 5 cm in diameter (a plastic Fresnel lens may be used). Mirrors M_1 and M_2 are to be aligned as shown in Fig. 98. The alignment is a bit time consuming because of the long distances involved.

A football field is a near perfect "optical bench" for measuring the speed of light. A regulation football field is about 120 meters long, measured diagonally from endzone corner to end-zone corner. When the three mirrors are properly aligned, both the "short" reference beam and the "long" one across the football field will be reflected into the photo-detector.

Next, the output of the photo-detector/amplifier (Metrologic 60-255 is a solid state photo-detector with a built in battery operated amplifier) is connected to an oscilloscope and the signal generator used to modulate the laser is turned on. The amplitudes of the "short" and the "long" beams must be balanced. This can be done by completely masking M_1 and M_3 in turn, while observing the oscilloscope. Then, the mirror which produces the most light is partially masked until the two waveforms on the scope are equal.

Starting at zero, the frequency of the signal generator is increased.

At low frequencies, the two beams are about in phase and they will interfere constructively. As the frequency is increased towards 600 kHz, the waveform on the oscilloscope will decrease in height until a nul is reached and only a horizontal line appears on the screen of the oscilloscope. This nul indicates that the "long" beam is arriving one-half wavelength after the "short" reference beam and thus destructive interference occurs. Knowing the frequency of the signal generator, it is easy to calculate the time needed to produce half a wavelength. By dividing the distance by the time, the speed of light then is obtained. Note: The "long" path is 240 meters longer than that of the "short" reference beam since the "long" path is the distance down-across and back-across the football field, thus 2 x 120 meters. The signal generator is set about 600 kHz to produce a nul. The period for the frequency is $1/(6 \times 10^{-5})$ sec, the travel time for a half wavelength is $1/(12 \times 10^{-5})$ sec. Dividing the distance by the time gives the speed of light as approximately 2.9×10^8 m/sec.

Figure 99 shows a variation of Fig. 98. The more powerful modulated laser (Metrologic ML-939) is used and the alignment procedure is simplified.

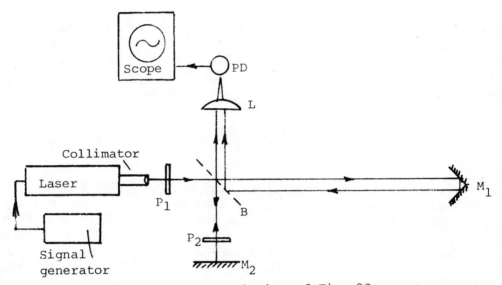

Fig. 99. A variation of Fig. 98.

The collimated laser beam is divided by beamsplitter B into two parts. The "short" reference beam is reflected to the front-surface mirror M_2 and then returned through B to the photo-detector. The "long" beam passes through B to the cube-corner etroreflector M_1 which returns the beam to the B to be reflected into the photo-detector. Polarizer P_1 is used to linearly polarize the beam emerging from the collimator. Polarizer P_2 is inserted into the "short" reference beam and adjusted to diminish its intensity relative to that of the "long" beam until the two beams are balanced.

References:
Education News from Metrologic, v2 nl p4 (Jan '72)
Page, D.N. and Geilker, C.D., Am J Phys v40 p86 (Jan '72)
Hinrichsen, P.F. and Crawford, J.C., Phys Teach p504 (Nov '75)
Knowles, C.H., Phys Teach p507 (Nov '75)

Methods for Spreading Laser Beams

Often it is necessary to spread a narrow, collimated laser beam to fill some aperture with light of uniform intensity. For example, this beam spreading is necessary to fill a diffraction grating to obtain the maximum resolving power; or, to provide a full field view in a Fizeau interferometer...

Reflection or refraction can be used for spreading laser beams. The mirror or lens used can be converging or diverging. Often astronomical or Galilean telescopes as beam expanders are used. The nonuniformity of the beam due to diffraction by dirt specks, scratches and by the aperture is a serious disadvantage when reflecting or refracting elements are used. Interference effects are also present because of the long coherence length and the narrow spectral width of the laser source.

Scattering is another method of spreading a laser beam. Most scatterers, however, are inefficient. Some scatter the light over a very large range of angles, others create "hot spots". In addition, the use of scattering with a laser source results in a granular far-field pattern. This granular (speckle) effect can be reduced if a laser beam expander is used to illuminate the scatterer. The speckle effect can be eliminated if the scatterer is rotated during the time of observation. Nematic liquid crystal cells operated in the dynamic scattering mode are employed also to eliminate or reduce the speckle effect.

Another spreader, coarse ground glass, spreads the laser beam at too great an angle.

Since most scatterers have hot spots, the best results often are obtained by combining refraction and scattering. A lens is used to produce a divergent beam so that the hot spot created by a weak scatterer is large enough to fill the desired angle or aperture. Good results can be obtained with "Tru-Site" which is a finely etched glass used as a non-glare glass in picture framing. (Tru-Site is a trademarked name of the Dearborn Glass Co., Argo, Illinois).

Reference:

Wiggins, T.A., Appl Opt v10 n4 p963 (Apr '71)

" " " " " " " " " " " " " "

Telescopes as Laser Beam Expanders and Collimators

Kepler's astronomical refracting telescope is shown in Fig. 100.

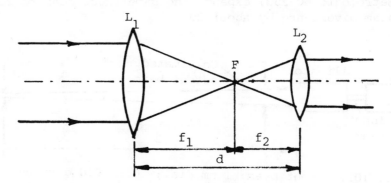

Fig. 100.   Astronomical telescope focused for
infinite object and image distance.

The incident rays are parallel at L_1 and come to focus at F. L_2 is located so that its first focus overlaps the second focus of L_1, and the rays diverging from the point F will leave L_2 parallel to each other. Thus, both the back and front focal lengths of a thin lens combination go to infinity when the two lenses are separated by a distance d equal to the sum of their focal lengths: $f_1 + f_2$. The astronomical telescope in this configuration is afocal, that is, without a focal length.

If the narrow laser beam is directed into the back end of this afocal telescope entering at L_2 and going from right to left, it will emerge at L_1 with an increased cross-section but still collimated. The angular magnification is given by $M = f_1/f_2$.

As long as $d = f_1 + f_2$, the telescope will be afocal even if L_2 is a negative lens. The telescope built by Galileo (Fig. 101) has such a positive/ negative lens combination.

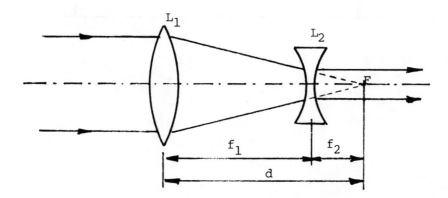

Fig. 101.   Galilean telescope.

In this case the focal length of the negative (L_2) lens must be numerically smaller than that of the positive (L_1) lens. Galilean telescopes are useful as laser beam expanders because they have no internal focal points

where a high power laser beam would ionize the surrounding air. Another
advantage of this type telescope is its small overall length.

Most collimators are focusing Galilean telescopes. An 8x collimator
(such as, for example, Metrologic 60-223) expands a 1 mm input laser beam to
8 mm and reduces the beam divergence by a factor of about 8. The 20x
collimator (Metrologic 60-253) expands the same input beam to 20 mm and
reduces the beam divergence by about 20.

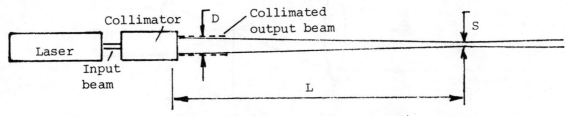

Fig. 102. Input beam: 1 mm dia.; 0.8 mRad divergence.

An expanded laser beam may be reduced to a smaller spot diameter S at
any given distance L (see Fig. 102). The table below compares the spot diameter (at a distance L) of a raw laser beam to that of an expanded and opti-
mally focused beam (at the same distance L), using the above mentioned 8x
and 20x collimators.

		Beam diameter in mm	
L in meters	Raw Beam	20x Collimator	8x Collimator
10	8	0.5	1.2
20	16	1.0	2.4
50	40	2.4	5.9
100	80	4.8	12.0
200	160	9.5	24.0
500	400	24.0	59.0

Lens-Pinhole Spatial Filter

In many experiments a clean, uncluttered laser beam is required for the recording or the observation of the fine details of interference and diffraction patterns. This is especially important for the needs of holography and optical data processing.

A lens-pinhole spatial filter is an instrument which transforms a collimated "noisy" laser beam into an expanding beam with a smoothly filtered spherical wavefront. The spatial filter can be used as the first element in a beam expander or merely to produce a "clean" diverging spherical wavefront. The instrument consists of a pinhole movable along the X and Y axes and a short focal length converging lens (often a standard microscope objective) movable along the Z axis. The power of the lens controls the divergence of the beam and the diameter of the pinhole aperture determines the degree of filtering. The lens focuses the input laser beam into the pinhole. The pinhole passes only the undisturbed portion of the laser beam and behaves as a low-pass filter.

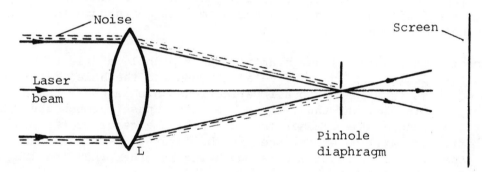

Fig. 103. Spatial filtering.

Set up the laser and its carrier on the optical bench and tighten the locking screw at the bottom of the carrier. Place a converging lens (25-50 mm focal length) in a component carrier and mount it securely on the optical bench a few centimeters in front of the laser. Aim the beam through the lens and observe the expanded, cluttered beam projected on the screen placed 1-3 meters from the lens. Repeat the experiment by using various converging and diverging lenses singly and in combination and observe the effect on the screen. Change the distance of the optical elements, rotate them around the Z axis (the direction of the laser beam) and note the rotation of the patterns on the screen. Some disturbances will remain stationary, others will rotate, or change in size, or both. The stationary disturbances originate in the laser itself, the others are added by the various optical elements in the path of the beam.

Next we shall align a lens-pinhole spatial filter to clean up the cluttered laser beam.

Alignment procedure:

1) Set up the laser and its carrier on the optical bench but do not

tighten the locking screw at the bottom of the carrier.

2) Place a converging lens (about 35 mm focal length) in a component carrier and mount it on the optical bench. Pass the laser beam through the center of the lens and fasten the locking screw of the lens carrier. Locate the lens about 5 cm in front of the output window of the laser.

3) Mount a pinhole (35 micron diameter or smaller) in a component carrier and locate the carrier on the bench at the approximate focal plane of the lens. Leave the locking screw loose.

4) The exact location of the focal point can be found by raising the laser from the optical bench and "panning" the beam at various parts of the lens, moving up or down and from side to side.

5) Adjust the lens-pinhole spacing until the movement of the small laser beam spot becomes stationary on the pinhole frame while panning north-south and east-west.

6) Once the pinhole is located in the exact focal plane of the lens, mount and secure the laser carrier on the optical bench.

7) By repeated vertical and lateral adjustments the pinhole can be located over the focal point and the laser beam will go through the hole and diverge on the other side.

By focusing the "noisy" laser beam through a pinhole a "clean" beam is obtained on the viewing screen. The metal surrounding the hole filters off the scattered beams from the various imperfections which are off the axis in the plane of the pinhole, and thus, the noise cannot pass through the hole. The smaller the pinhole, the cleaner the beam will emerge after the pinhole but more difficulty will be experienced in the laser-lens-pinhole alignment procedure. Spatial filtering of lasers operating in other than the TEM_{oo} mode is extremely inefficient.

To avoid the tedious adjustment procedure involved in obtaining a clean laser beam, an inexpensive Metrologic 60-618 Spatial Filter was used in many of the experiments.

The 60-618 Spatial Filter may be mounted on an optical bench by means of the 13.7 mm pin provided, or it may be mounted directly to a laser by means of the threaded end cap and mounting plate equipped with a standard 1"-32 TPI thread. Alignment is simpler and more reliable if the Spatial Filter is mounted directly to the laser.

Visual adjustment:

1) Place a sheet of white paper about 15 cm in front of the Spatial Filter for a viewing screen.

2) Loosen locking ring B (see Fig. 104) and turn it counter-clockwise until it is a quarter inch or more away from the Spatial Filter body.

3) Simultaneously, adjust the two lens adjusting screws until a spot of light appears on the viewing screen. When the spot appears, adjust each lens

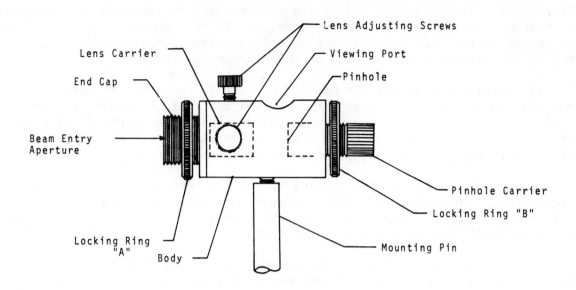

Fig. 104. The lens-pinhole spatial filter.

adjusting screw alternately until the spot is of maximum brightness and roundness.

4) If the bright spot is surrounded by one or more rings, the pinhole is improperly focused. If this is the case, continue with Step #5. However, if no rings are evident and the spot is of comparable intensity to the beam entering the filter, then the pinhole is properly focused and one may go ahead to Step #7

5) Rotate the pinhole carrier in either direction until the spot is again visible on the screen. If the spot is noticeably brighter, or if the number of surrounding rings has decreased, then the direction of rotation is correct. If the spot is of less intensity, or if the number of rings has increased, then the rotational direction is incorrect and must be reversed.

6) Rotate the pinhole carrier in the correct direction, observing the spot as it reappears with each revolution. Continue this procedure until the spot appears without surrounding rings and is approximately as bright as the beam entering the Spatial Filter.

7) Hold the pinhole carrier in place with one hand while rotating locking ring B clockwise with the other hand until the locking ring is snug against the spatial filter body.

8) Alternately, adjust the lens adjusting screws for optimum brightness and roundness of the spot.

Adjustment with Photometer:

1) Go through steps #1 to #6, as described above, for optimum visual adjustment.

2) Measure the intensity of the beam out of the Spatial Filter with a photometer, such as the Metrologic 60-230.

3)   Rotate the pinhole carrier one-quarter turn in either direction, and alternately adjust the lens adjusting screws until the maximum beam intensity is obtained, as monitored by the photometer. If this output is greater than the output measured in Step #2, the rotational direction is correct. If not, reverse the direction of rotation of the pinhole carrier.

4)   Continue to adjust the pinhole carrier in one-quarter turn steps, each time adjusting the lens adjusting screws for maximum reading of the photometer. Repeat the procedure until any further adjustment of the pinhole results in a lower maximum reading than the preceding adjustment. The pinhole is now properly focused and the Spatial Filter is ready for use.

The pinhole size in the 60-618 Spatial Filter has been optimized for light through-put. For contact printing copies of holograms and other critical applications, the 35 micron pinhole may be somewhat large and a smaller (25 micron or less) pinhole diameter is required.

The size of the pinhole is given by $D = 2\lambda f$, where D is the diameter of the pinhole, λ is the wavelength of the laser light, and f is the focal length of the lens divided by the diameter of the laser beam at the lens.

A useful technique for placing a pinhole in the focal plane of the lens is to move the pinhole back and forth until the size of the reflected laser speckle is largest. The reason being that the laser beam is smallest in the focal plane and the smallest spot on the pinhole produces the largest speckle. The micropositioner is then used to center the pinhole in the focal plane of the lens.

_____

For further reading:

B.J. Pernick, Rev Sci Instrum v45 n11 p1344 (Nov '74)

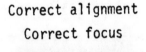

Correct alignment
Correct focus

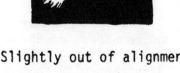

Good alignment   Slightly out of alignment
Out of focus   Good focus

Fig. 105.   Correct and incorrect
alignment.

Measuring the Concentration of Colloidal Solutions or the Dirt Content of Water

First a simple apparatus is described for measuring the concentration of a colloidal solution.

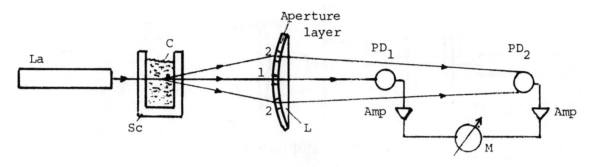

Fig. 106

As shown in Fig. 106, the direct laser beam is incident on the colloidal solution (C) within the sample cell (Sc) made of glass. A ray shielding layer is formed on the surface of a lens (L). At the center portion of the ray shielding layer a small circular aperture (1) about the diameter of the laser beam, and a ring shaped thin slit (2) concentrically surrounding the center aperture are provided. A photo-detector (PD_1) is provided on the axis of the direct laser beam, and a similar photo-detector (PD_2) is provided at the imaging position of the solution (C). The outputs of these light sensitive elements are amplified (Amp) and fed to a ratio meter (M) such as a potentiometer type or wattmeter type device.

When a colloid is not contained in the sample cell, the laser beam passes through the cell, the center aperture of the lens, and enters PD_1 only. The output comes from PD_1 since PD_2 does not receive any scattered light. If a colloid is contained in the sample cell, a portion of the scattered light is focused by the lens L on PD_2. Accordingly, an output is obtained from PD_2 and, at the same time, the output of the light sensitive element PD_1 decreases. The ratio of the outputs of the light sensitive elements PD_1 and PD_2 at this time corresponds to the concentration of the colloidal solution.

The ray shielding layer may be spaced away from the surface of the lens. It may be replaced by a similar aperture disk provided in front of or in the rear of the lens L. Fig. 107 shows three configurations.

An accurate instrument for measuring turbidity (the dirt content of water) has been developed by a group of undergraduates at the Institute of Optics, The University of Rochester. They built an instrument of the ratio type in which an error in one part of the system is compensated for, keeping the ratio constant. The device is called a dual-beam turbidimeter and its arrangement is shown in Fig. 108.

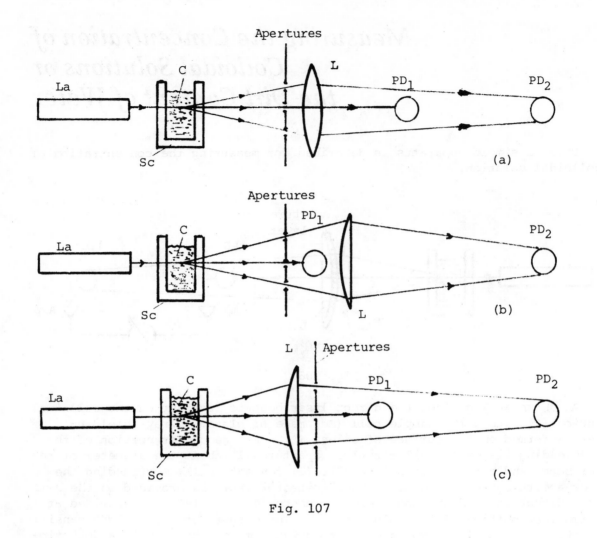

Fig. 107

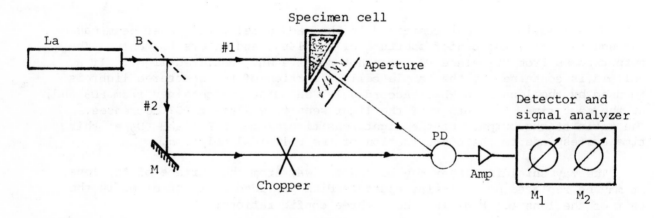

Fig. 108

The light from the laser is split into two beams at the beam splitter (B). The transmitted portion (#1) enters the specimen cell (a hollow prism made of glass). The refracted rays pass through an aperture and fall on a photo-detector (photo-diode; photo-transistor). The intensity of these scattered rays varies with the turbidity of the specimen.

The beam reflected (#2) by the beam splitter falls on a front surface mirror from which it is directed onto the photo-detector. Test beam (#1) and reference beam (#2) combine in the photo-detector where they are added algebraically. The reference beam (#2) is interrupted by a motor-driven chopper or an oscillating chopper. The photo-detector output is amplified and a circuit separates the output into two parts: a DC signal and an AC signal. The DC signal is the sum of the current generated by the test beam transmitted through the specimen plus the DC component of the chopped reference beam. It is measured by M_1. The amplitude of the AC signal is proportional to the intensity of the reference beam, and is measured by M_2.

The easiest way to use the instrument is to fill the specimen cell with clear water and record the ratio of the two meter readings: M_1/M_2 (water). The clear water is then replaced by the polluted water and a second set of readings is taken: M_1/M_2 (specimen). The transmission of the specimen, as calibrated against clear water, is expressed by the ratio: M_1/M_2 (specimen)/ M_1/M_2 (water). An experimental check can be made with a specimen prepared by suspending polystyrene spheres (2 microns in diameter) in distilled water. The test is based on the fact that concentrations of spheres of uniform diameter attenuate light by a predictable amount. Spheres of this kind are available commercially (Duke Standards Company, 445 Sherman Ave., Palo Alto, California 94306) and come in the form of concentrated solution.

The elements of the turbidimeter, the electronic circuitry and the calibration technique are all described in detail in the referenced article (*).

References:

T. Hasegawa, U.S. Pat. No., 3,310,680 (21 Mch '67)
(*) C.L. Stong, Sci Amer v228 p112 (Jne '73)
A.N. Sakharov and K.S. Shifrin, Opt. Spectrosc v39 p208 (1975)
T.N. Padical, J.G. Kereiakes and P.J. Robbins, Med Phys v3 169 (Mch/Apr'76)

" " " " " " " " " " "

Simple Polarization Demonstrations

The Lazy Susan setup used in the following demonstrations is the same as the one used before for ray-optics exercises. The direct beam from a low-power He-Ne laser is passed through a short-focus cylinder lens (a glass or plastic rod of 2-4 mm diameter). The transparent rod is mounted on, or near, the laser housing where the beam exits from the laser. The light emerging from the rod is fan shaped. The rod is oriented horizontally and this results in a thin vertical sheet of light which cuts across the horizontal plane of the Lazy Susan table, producing a bright ray streak on the table. Between the laser and the Lazy Susan table there is a short (25 cm) optical bench to carry polarizers and retardation plates.

1. Polarization of laser light.

All laser light is polarized. The total output power of a so-called "linearly polarized laser" lies in a single, defined   plane of polarization. The output power of a so-called "randomly polarized laser" is distributed between two orthogonal planes of polarization.

Internal mirror lasers are different from Brewster window type lasers in that there is no polarizing element within the optics cavity. The output beam of this type laser consists of two or more colinear beams of linearly polarized light of slightly different wavelengths. The optical power associated with each polarization plane will shift over time. The power shifts between the two orthogonal polarization components of the beam can be quite significant.

The phenomenon described can be observed by placing a linear sheet polarizer in the beam path between the cylinder lens and the Lazy Susan table (see Fig. 109).

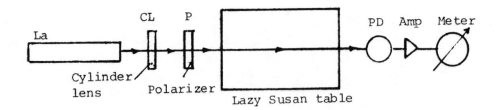

Fig. 109

The laser used is an internal mirror type, having no Brewster windows. The emitted light is having an unspecified orthogonal polarization. The linear sheet polarizer is inserted in the beam path and the intensity of the ray streak on the Lazy Susan table is observed over a period of time. The intensity fluctuations may be measured with a photo-detector, amplifier, and ammeter and charted as a function of time. First, rotate the polarizer until the brightest beam is transmitted through the filter. At this position the marking on the polarizing filter (indicating the easy transmission axis of the polarizer) indicates the direction of polarization of the laser beam.

Leave the filter in this position and monitor the photometer for several minutes and record its reading at intervals of ten seconds. Make a graph of beam intensity versus time. Examine your graph to determine whether or not there is any periodicity in the variation of polarization planes.

To determine the percentage polarization of the laser beam, rotate the polarizer and take two meter readings: Let I_{max} and I_{min} represent the maximum and minimum meter readings. The percentage polarization of the laser beam is defined as:

$$\text{Percent polarization} = \frac{I_{max} - I_{min}}{I_{max} + I_{min}} \times 100$$

2. Control of Brightness.

Two linear polarizers in the path of the laser beam may be used as a light valve (or light dimming device) by rotating one against the other one in fixed position (Fig. 110).

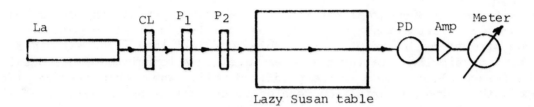

Fig. 110

Linear polarizers P_1 and P_2 are placed in the indicated positions. One is in fixed position while the other one can be rotated. If two linear polarizers are placed in series and aligned with their polarizing axes parallel, maximum total transmission will result. If one polarizer is then rotated 90 degrees with respect the other, minimum transmission will result. The brightness of the transmitted beam falls rapidly as the polarizers near the "crossed" position.

The "Law of Malus" states that the amount of light transmitted varies as the square of the cosine of the angle between the axes of the two polarizers:

$$I = I_o \cos^2 \phi$$

where: I = transmitted intensity,
        I_o = maximum intensity of light impinging on the second polarizer,
        ϕ = the included angle between the axes of the two polarizers.

Note: In order to obtain correct readings with the photo-detector, the experimental setup must be carefully masked against stray light.

The transmission axes of P_1 and P_2 are set parallel to each other to obtain maximum deflection of M. P_2 is then turned through 30, 45, and 60 degrees; the meter deflection is reduced to 3/4, 1/2, and 1/4 of its original reading. In a 360-degree movement of the rotatable polarizer there will be two positions of maximum and two positions of minimum transmission.

3. <u>The Effects of a Half-wave Retardation Plate on Polarized Light.</u>

Many crystals, some plastics, and some other materials have different optical properties in different directions. This dependence of optical properties on direction is known as birefringence, and such materials are said to be doubly refracting.

When an incident light ray is passed through a doubly refracting material whose optical axis is oriented at other than 0 or 90 degrees with respect to the interface surface, the incident ray is split into two polarized components at right angles to each other. The indices of refraction are different in a birefringent material for two mutually perpendicular directions of polarization. Thus, the two rays traveling through a doubly refracting material do so at different velocities. When the two rays emerge from the substance they recombine. Since one ray generally lags behind the other, they recombine out of phase. A so-called "retardation plate" which causes one ray to lag behind the other by one-half a wavelength is a "half-wave plate." The extent of retardation depends on the nature of the doubly refracting substance, its thickness and the wavelength of light involved.

A "half-wave" (180-degree) retarder can be used to change the direction of the plane of polarization. The output of many lasers is plane polarized in a given direction. It is often desirable to rotate this fixed azimuth at will without physically rotating the laser itself. A technique of doing this uses a half-wave plate placed in the polarized beam. When the half-wave plate is rotated, the output polarization rotates twice as fast as the plate. The effect is demonstrated by the setup shown in Fig. 111.

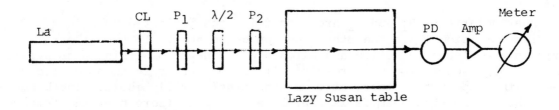

Fig. 111

Polarizer P_1 is placed in the beam with its transmission axis horizontal. Polarizer P_2 is crossed, that is, its transmission axis is oriented vertically. None of the light which emerges from the first polarizer can pass through the second polarizer. The half-wave plate is placed between the crossed polarizers. (Two quarter-wave plates may be used in series with their axes parallel to make a half-wave plate.) Note that the ray streak on the Lazy Susan table is brightness modulated. In a complete 360-degree rotation there will be four maxima and four minima as the half-wave plate is rotated. Also note that the maxima are seen when the angle between the polarizer axes and the wave plate axis is 45-degrees (also at 135, 225 and 315 degrees). Thus, when the wave plate is at 45-degrees, the plane of polarization of P_1 is rotated through $2 \times 45° = 90°$, and thus the light is transmitted by P_2.

Commercial retardation plates are usually made of high quality crystalline quartz and thin sheets of mica sealed between glass plates. Fairly good retardation plates can be improvised with cellophane.

Slip a piece of cellophane (from a cigarette package) between the crossed polarizers. Now light gets through! Rotate the piece of cellophane through 360 degrees and observe the maxima and minima. The effect will probably be not as clear-cut as with commercial half-wave plates designed for the laser wavelength. You can also try to use cellophane tape ("Scotch"-brand transparent tape) to make a half-wave plate. Stick one layer of clear cellophane tape on a slide cover glass or microscope slide for mechanical support. It is possible to construct a half-wave plate out of two identical pieces of cellophane neither of which is quite the right thickness to give 180-degree retardation for the laser wavelength. The two sheets are placed in series and they are rotated, "tuned," relative to each other until exactly 180-degree phase difference is achieved.

Household plastics like Saran Wrap or Handi Wrap (used to wrap sandwiches) show very little effect between crossed polarizers. By stretching the plastics 45-degree to the axes of the crossed polarizers a strong effect will be observed because the long organic molecules of the plastic tend to get aligned by the stretching. Seven or eight layers will make a fairly effective quarter-wave plate. A Saran Wrap half-wave plate will take about 14-16 layers. To "tune" the retarder add or substract one layer. Tape the plastic layers to a cardboard which has a hole cut in it to serve as a window for the laser beam.

4. The Quarter-wave Retardation Plate and Circularly Polarized Light.

A beam of light passing through a doubly refracting crystal is split into two components, each linearly polarized, but with their axes of polarization mutually perpendicular. The two components do not travel at the same speeds. If a coherent laser beam is directed into the crystal, the two components are out of phase when they emerge from the crystal. A retardation plate which causes one wave to drop behind the other by just one-quarter of a wavelength is called a "quarter-wave plate."

Quarter-wave plates find their chief use in changing elliptically or circularly polarized light to plane-polarized light and to convert linearly polarized light to circularly polarized light. They are also used in investigating the nature of a beam of light.

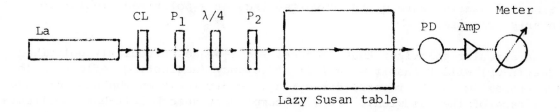

Fig. 112

Figure 112 shows the setup for the investigation of quarter-wave plates. At the beginning of the exercise the second polarizer, P_2, is not on the optical bench as yet! The light emerging from the linear polarizer, P_1, can be given a "twist" by inserting a quarter-wave plate into the polarized beam.

The vibrations are no longer confined to a single plane, but instead form a helix. If a linearly polarized coherent beam is directed at the quarter-wave plate with the plane of polarization of the beam at 45-degrees angle to the optical axis of the retarder, the input beam will be divided into two components. The beam leaving the retarder will consist of two linearly polarized components whose planes of polarization are mutually perpendicular and whose phase difference will be 90 degrees. The output from the retarder can be referred to as circularly polarized light. Here the direction of vibration rotates around the direction of propagation in a corkscrew fashion. When the second polarizer (P_2 = analyzer) is placed on the optical bench in the path of the beam, the light transmitted through P_2 does not change in intensity as P_2 is rotated.

The quarter-wave plate must be designed for use with a specific source of light. A quarter-wave plate designed for green light will not be satisfactory for the red light of a He-Ne laser. Unless the quarter-wave retarder is designed for the 633 nanometer wavelength of the laser beam, the resultant beam will be elliptically polarized to varying degrees.

A circular polarizer is a "sandwich" consisting of a sheet of linear polarizer bonded to a quarter-wave retardation sheet oriented at an angle of 45 degrees to the transmission axis of the polarizer. The polarizer and the retarder can be bonded together with the axis of the retarder placed at 45 degrees to the right or left of the polarizer's transmission axis. If to the right, the circularly polarized light is a right-handed helix (clockwise rotation); if to the left, it is a left-handed helix (counterclockwise rotation).

Referring again to Fig. 112, remove the quarter-wave plate. Set P_2 for extinction. Replace the quarter-wave plate between P_1 and P_2, its axis set at any angle except 45 degrees relative to the axes of the polarizers. When the analyzer, P_2, is rotated periodic maxima and minima of illumination appear on the Lazy Susan table. These are due to elliptical polarization and none of the positions of P_2 will produce total extinction. Re-set the quarter-wave plate at 45 degrees to change the plane polarized to circularly polarized light. Rotating P_2 will produce no intensity change in the ray streak observed on the Lazy Susan table.

Next, introduce a second quarter-wave plate between the one already on the optical bench and P_2. When you rotate P_2 you will observe that in effect the second quarter-wave plate converts circularly polarized light into linearly polarized light.

A circular polarizer that produces right-handed circularly polarized light (RHCP) will transmit with 100% efficiency (neglecting reflection losses) RHCP traveling in the reverse direction (i.e. incident on the quarter-wave plate face of the sandwich). It will completely absorb left-handed circularly polarized light (LHCP) incident on the quarter-wave plate face. Further, when a RHCP beam is reflected from a non-depolarizing (specular) surface, it becomes a LHCP beam, and thus it cannot pass through the same filter that originally polarized it. These phenomena can be demonstrated with the setups shown in Figs. 113, 114 and 115.

Referring to Fig. 113, light passing through the circular polarizer, CP, emerges, let us say, as a right-handed beam. After it is reflected from

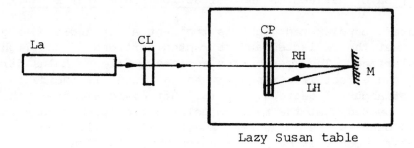

Lazy Susan table

Fig. 113

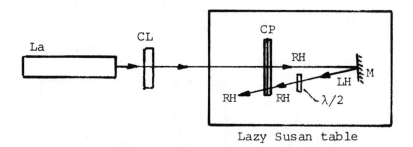

Lazy Susan table

Fig. 114

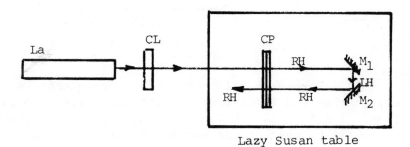

Lazy Susan table

Fig. 115

mirror, M, it becomes left-handed and thus the reflected beam cannot pass
through CP. It is blocked.

Fig. 114 shows the same arrangement as in Fig. 113 except that a half-
wave plate is inserted in the beam reflected by the mirror. The half-wave
plate converts the left-handed circularly polarized light into right-handed
circularly polarized light and now the beam goes through CP.

Fig. 115 shows a setup involving reflections from two specular surfaces.
The beam emerging from CP is right-handed. The sense of rotation reverses
with reflection on M_1. When this left-handed beam is reflected by M_2, it is
reversed again and the resulting right-handed beam will pass through CP on
the return path.

5. Polarization by Reflection and Refraction - Brewster's Law.

If natural, unpolarized light is incident at an angle on a glass plate, it is found that the reflected and refracted portions of the light become partially polarized. The French army engineer, Étienne Malus first observed in 1808 that the effect is strongest when the light is incident on the glass at an angle of about 57 degrees. In 1812 Sir David Brewster, the Scottish physicist discovered that the maximum polarization occurs when the reflected and refracted bundles are 90 degrees apart. Figure 116 shows the optical arrangement for demonstrating polarization by reflection.

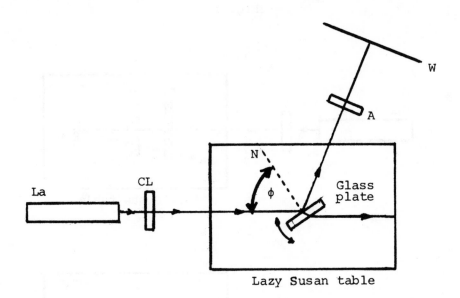

Fig. 116

The plate glass (microscope slide or slide cover glass) is placed on the Lazy Susan table so that the laser light is incident at the so called "polarizing angle" (ϕ=57°). As the analyzer, A, a sheet polarizer is rotated about its axis, the light seen on the screen, W, will alternately become bright and dark.

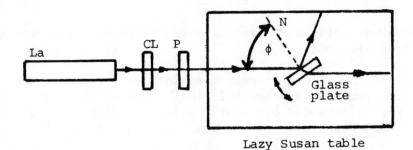

Fig. 117

A glass plate may be used either to produce a beam of plane-polarized light (used as a polarizer, as shown in Fig. 116), or to detect the presence of polarization in a beam (used as an analyzer, as shown above in Fig. 117).

A sheet polarizer, P, is placed in the path of the laser beam with its transmission axis oriented horizontally. At the outset of the exercise, the glass plate is set about 45 degrees, the reflected ray streak is clearly seen on the Lazy Susan table. As the glass plate is turned clockwise to increase the angle of incidence, the reflected light gradually decreases in intensity. When φ is approximately 57°, the reflected component will be minimum. This is Brewster's angle at which the refracted light that passes through the glass plate is exactly 90° to the reflected component. According to Brewster's Law, tan φ = n(glass)/n(air); thus the "polarizing angle" is a function of the index of the transmitting medium and the wavelength of the incident light.

After Brewster's discovery Malus recognized that not only the reflected but also the refracted (transmitted) light becomes polarized. The reflected beam, although completely polarized, is weak; the transmitted beam, although strong, is only partially polarized. The device called "pile-of-plates polarizer" was invented by Dominique Arago in 1812. With two dozen microscope slides or slide cover glasses held together with rubber bands we can improvise a primitive but quite effective polarizer of this sort (see Fig. 118).

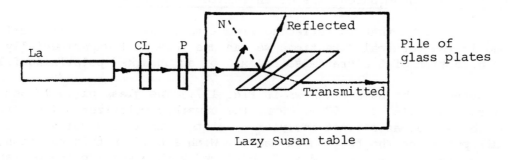

Fig. 118

Two dozen thin glass plates will give better than 80% polarization in the transmitted portion of the light. The laser beam is incident at the polarizing angle on the first surface of the pile of plates. The reflected and refracted rays are polarized in planes perpendicular to each other. The arrangement shown in Fig. 118 uses the pile-of-plates device as an analyzer. The laser beam is polarized with sheet polarizer, P. If the plane polarization lies within the plane of the Lazy Susan table (horizontal polarization axis), only the transmitted ray streak will be seen. As the polarizer is rotated in its plane the transmitted light diminishes and the reflected ray streak appears. The intensity of the reflected ray streak rises to a maximum when the polarizer's axis is vertical. At this point the transmitted beam is at a minimum.

One interesting application of Brewster's angle is in the design of a glass window having approximately 100% transmission, called a Brewster Window. By applying Brewster windows on laser plasma tubes between the mirrors constituting the resonator for the laser energy, an output beam with single linear polarization is produced.

Internal mirror lasers are different from Brewster window type lasers in that there is no polarizing element within the optics cavity. Light emerging from lasers having no Brewster windows have a polarization which may change as a function of time. The effect can be disturbing in split beam applications, such as interferometry and holography, where a coated glass beam splitter is employed. The effect may be overcome by decreasing the angle of the beam-splitter to the incident beam. If the angle of incidence is less than 20°, the effect is reduced considerably (see Fig. 119).

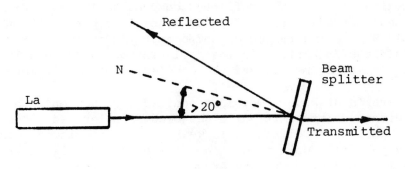

Fig. 119

6. Finding the Transmission Axis of an Unmarked Polarizer.

Since the light that is reflected from a vertical non-metallic surface is polarized in the vertical direction, we can use this fact experimentally to find the direction of the transmission axis of an unmarked polarizing filter.

The setup is the same as shown in Fig. 117. The glass plate is set at the "polarizing angle" ($\phi = 57$). P is the unmarked polarizer. The vertical glass plate is used as a fixed analyzer. Rotate P until the reflected ray streak disappears on the Lazy Susan table. With P held in this position, mark the polarizer at "9 o'clock" and at "3 o'clock" with a small piece of tape or dots of nail polish. Once marked, you will know in the future that this filter's axis of polarization is horizontal when the marks are at the sides and vertical when they are at the top and bottom.

References:

A.S. Marathay, Optical Engineering v14 pS-17 (Jan/Feb '75)
A.S. Marathay, Optical Engineering v14 pS-56 (Mch/Apr '75)
E.S. Hass, Laser Tech Bull #7 Spectra-Physics, Inc. (1975)
Application Note: AN-970-1, Metrologic Instruments, Inc. (1970)
"Polarized Light," Polaroid Corporation.
J.G. Winans, Am J Phys v21 p170 (Mch '53)
S.D. Cloud, Am J Phys v41 p1184 (Oct '73)

" " " " " " " " " "

Scattering of Light and Polarization

The polarization of the light from an unpolarized beam that is scattered by a suspension of fine particles can be demonstrated with any one of the arrangements shown in Fig. 120.

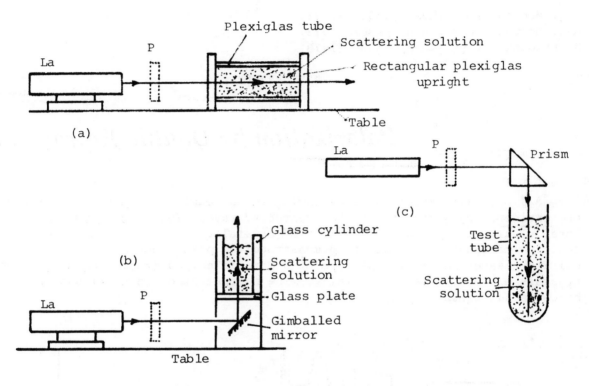

Fig. 120. Polarization by scattering (side view).

The polarization of light by scattering was first observed by Lord Rayleigh in 1871 using colloidal sulphur as the scattering material. Colloidal sulphur is formed by adding acetic acid to a weak solution of photographic fixer (sodium thiosulphate). A weak colloidal suspension of $AgNO_3$ or $AuNO_3$ in water is also a suitable scattering medium for observing Rayleigh scattering, which requires that the scattering particles be small compared to the optical wavelength. A convenient way to produce a colloidal suspension is by stirring into the water a very small amount of instant nonfat dry milk.

The transparent vessel containing the scattering material is placed in the laser beam path. The observer views the light scattered at right angles to the beam. Scattered light will be seen all around the transparent vessel. If the observer looks into the side of the vessel through a polarizing filter, the suspension will seem brighter or darker as the polarizer is rotated. Note: The laser used does not have a plane polarized output and the polarizer, P, shown in the drawing is not in the beam path!) When observed through a hand-held polarizer, the laser beam passing through the suspension will seem brightest when the polarization axis of the filter is at right angles to the laser beam. With the polarizer P in the beam path as shown in the drawings and observing directly (without the hand-held polarizing filter), the scattered light is brighter in the direction perpendicular to the plane of polarization and very faint or absent in the direction parallel to it. If the polarizer P in the beam path is rotated, these directions of maximum and minimum intensity will rotate with it.

Caution: Do not look directly into the laser beam!

_____

References:

W.S. Moore,   Am J Phys   p1536   (1971)
M.H. Moore,   Phys Teach p436   (1973)
H. Kruglak,   Phys Teach p550   (1973)

"""""""""""

Polarization by Double Refraction

All crystals except those belonging to the cubic system are birefringent. One incident ray gives rise to two refracted rays. These two rays are separated clearly in only a small number of crystals. Crystals of calcite (calcium carbonate) are used to demonstrate the phenomenon of double refraction. Calcite cleaves easily in the form of a parallelepiped (rhombohedron). Such crystals are available commercially. The demonstration setup illustrated below is not drawn to scale.

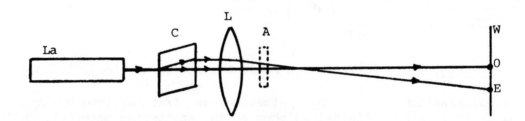

Fig. 121. Demonstration of double refraction.

The crystal (C) is mounted in a rotatable holder so that its smallest surface is in the plane of rotation. The laser beam is incident normally on this surface. The input beam is split at the entrance face of the crystal and then leave the exit face at two points. These two rays are visible as they pass through the crystal. The non-deviated ray is the "ordinary" ray (O). The other ray is deviated and is called the "extraordinary" ray (E). The emergent spots may be magnified by a short focal length convex lens (L),(36 mm f.l.), and projected on a screen or distant wall (W).

The refractive index of calcite for the O-ray is about 1.66 and for the E-ray it varies from 1.66 to 1.49 depending on its direction of travel relative the optic axis of the crystal. If the calcite crystal is turned about an axis normal to the entrance face, the extraordinary ray (E) is seen to rotate around the ordinary ray (O) which remains fixed.

Any unpolarized or linearly polarized light ray passing through a calcite crystal is resolved into components linearly polarized at right angles to each other. The E-ray vibrates in the principal section of the crystal and the O-ray in the perpendicular plane. A sheet polarizer (A = analyzer) placed after the lens will extinguish one or the other of the images on the screen, depending on the orientation of the polarizing axis. Thus, the plane of polarization of each ray can be determined.

The same exercise can be repeated with a parallelepiped of iodic acid, a biaxial crystal. If this crystal is rotated around the laser beam direction, the two rays describe cylinders.

"""""""""""

Projecting Crystal Patterns in Convergent Light

Polarized light is used extensively in the field of mineralogy. The patterns produced by transparent specimens cut into a thin plate, placed between two polarizers, and illuminated with monochromatic convergent light, are instructive when making comparisons of the symmetries of different types of crystals. The patterns projected (or photographed) indicate whether the crystal is (a) uniaxial or biaxial, (b) if uniaxial, where the optic axis is, and (c) if biaxial, the angle between the optic axes.

A simple arrangement for projecting crystal patterns on a nearby screen is illustrated in Fig. 122.

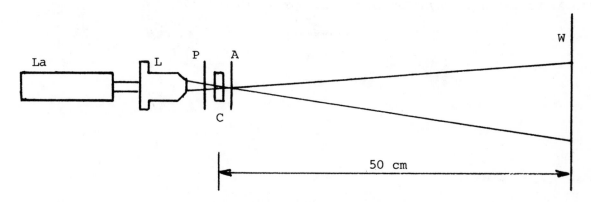

Fig. 122. Crystal Pattern Projector.

Light is converged on the crystal (C) by a short-focus lens (L), such as a 20x - 30x microscope objective. The crystal is sandwiched between sheet polarizers (P and A). The screen, W, must be fairly close to the "sandwich." This screen distance is important when a biaxial crystal (such as muscovite mica) is observed because the distance between the "eyes" in the pattern is almost one and one-half times the distance from the lens to the screen.

To get good results with high contrast, the polarizing plates must be either parallel or crossed and the axes of the crystal plate must be at 45° to the axes of polarizers P and A. Using calcite, quartz, mica, sugar and aragonite crystals many interesting patterns can be obtained between parallel and crossed polarizers and circular polarizers (see *).

If a uniaxial crystal (calcite, for example) is cut perpendicular to the optical axis, one observes concentric rings centered along the optical axis and a black cross is superimposed on the rings. The two branches of the cross (called "brushes") are parallel and perpendicular to the transmission axes of the polarizers. When circularly polarized light is used, the cross may be eliminated.

If a uniaxial crystal plate   cut parallel to its optical axis is observed, the interference fringes are hyperbolas with axes parallel and perpendicular to the optical axis. The center of the hyperbolas is dark or bright depending on whether the polarizers are crossed or parallel. Biaxial crystals show much more complicated figures of interference.

It is well known that liquid crystals exhibit birefringence. Many liquid crystals are available commercially. The room-temperature nematic liquid crystal methoxy-benzylidene-n-butylaniline (MBBA) is well suited for these experiments. A thin layer of MBBA between glass slides is placed between crossed polarizers. The converging laser beam displays the orientation of the optic axis on the screen. The value of $\Delta n = n_o - n_e$ can be obtained from the angular spacing of the fringes (see **).

References:

E.H. Wood, "Crystals and Light," Van Nostrand, New York, pp94-100
*S.H. Burns and M.A. Jeppesen, Am J Phys v28 p774 (Dec '60)
**D.A. Balzarini, Phys Rev Lett v25 p914 (5 Oct '70)

"""""""""""

Optics Experiments with Liquid Crystals

Liquid crystals were first identified at least ninety years ago, but it has been only during the last fifteen, or so, years that their unique physical properties and technical applications received any real attention.

Liquid crystalline materials have characteristics which lie between those of a true liquid and a true crystal. They have, for example, the relatively low viscosity of a liquid along with some of the optical properties of a crystal, such as molecular ordering.

The liquid crystal mesophase of an organic material is a state of matter existing between the solid and the isotropic liquid phases.

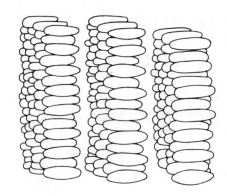

(a) Smectic phase

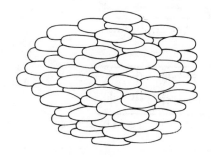

(b) Nematic phase

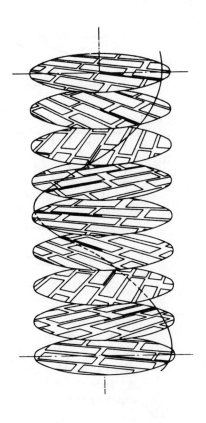

(c) Cholesteric phase

Fig. 123

Liquid crystals are divided conveniently into three classes, - smectic, nematic, and cholesteric. In smectic liquid crystals, the molecules are arranged in layers with their long axes approximately normal to the plane of the layers. The planes can slide easily over each other (Fig. 123-a). In nematic substances, the arrangement of molecules is analogous to a long box filled with round pencils which can roll around and slide back and forth, but remain parallel or nearly parallel to one another in the direction of the long axis of the pencil (Fig. 123-b). In cholesteric materials, the molecules are arranged in parallel layers. These layers are very thin with the long axis of the molecules parallel to the plane of the layers. The direction of the long axis of a molecule is slightly and systematically rotated from the direction of the axis of the molecules in adjacent planes. The result is a helical structure which shows great optical activity because the layer distance of several thousand Angstroms corresponds to optical wavelengths (Fig. 123-c).

Liquid crystal cell construction.

The study of optical effects in liquid crystals became very productive when it was discovered that thin films (5-25 μm) yielded a wealth of optical information not obtainable from bulk studies. Thin films show many different textures depending on the nature of the liquid crystal, the surfaces of the substrate and cover slide, and the temperature. Figure 124 shows some of the ways in which liquid crystal molecules may lie in relation to the boundary surfaces.

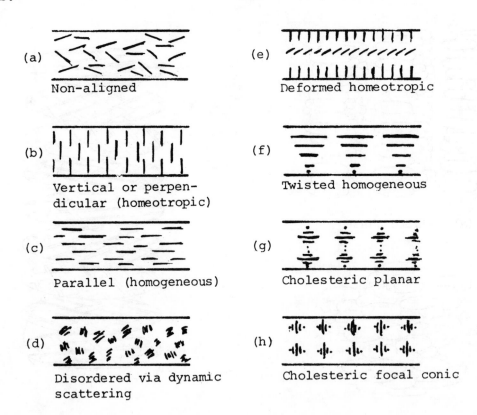

Fig. 124. Arrangements of liquid crystal molecules in thin film layers.

The liquid crystal material can be examined by containing it between two plane-parallel plates such as Kodak slide cover glasses. Electro-optical effects are investigated by placing a thin Mylar or Teflon spacer between two pieces of 'conductive glass' which are offset about 5-10 mm (Fig. 125). The assembly can be clamped together, or the sandwich cell can be bonded with epoxy cement.

NESA Glass (TM) is a transparent, electrically conductive, tin-oxide coated glass with a typical normal surface resistivity range of 60 to 300 ohm/square. Due to the nature of the coating process, NESA Glass is not as flat as regular termally tempered glass. The flatness and optical distortion of NESATRON Glass (TM) is as good as that of the substrate because the coating is applied at temperatures well below the softening point of the glass and flatness and anneal, or temper, are not affected. (Both NESA and NESATRON Glass are manufactured by PPG Industries, Industrial Glass Products, Pittsburgh, PA 15222). For both types the index of refraction of the glass substrate is 1.52 and the conductive coating 2.0. The conductive coating is essentially as hard as the glass substrate. The coated surface can ususally be identified by its iridescence when viewed in reflected light. Positive identification can be made by checking with an ohmmeter. The thickness of the coating on NESATRON Glass for surface resistivities between 15-1,000 ohm/square ranges from approximately 300 to 2,700 Angstroms.

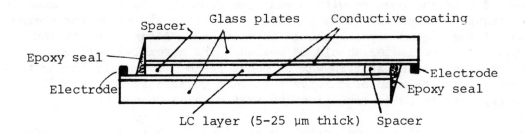

Fig. 125. Liquid crystal sandwich cell (not drawn to scale).

It is necessary to clean the glass plates before they are used in a LC cell. This needs to be done immediately prior to use, as storage after cleaning can only result in recontamination. Abrasive cleaners should not be used on either surface of the conductive glass. The glass plates should be given a 10 to 15 minute wash in a commercial detergent. (The use of an ultrasonic cleaner filled with a commercial degassing detergent is recommended.) The glass plates are then rinsed in water and immersed in 1,1,1 Trichloroethane, or Trichloroethylene, or Freon TF for ten minutes. Next, a rinse in distilled water is applied and then the plates are dried with acetone or methanol. Drying should be completed in the cleanest possible atmosphere.

In certain device applications it is desirable to provide a nematic liquid crystal film in a homogeneous texture, that is, large areas in which the LC optic axis is oriented parallel to the boundary surfaces which enclose the nematic layer. Homogeneity can be achieved by rubbing the boundary surfaces. The glass slides are rubbed by moving them back and forth about two dozen times with even pressure on a flat surface covered with filter paper or cheesecloth. The molecules of the nematic liquid crystal become oriented in the rubbing direction to form the equivalent of a slab of a uniaxial crystal with the optic axis lying in the rubbing direction. The assembled LC cell may be heated to about 50°C to achieve alignment within a few minutes. At room temperature alignment may take several hours.

It is often desirable for device applications to provide a nematic layer in a homeotropic texture, that is, large areas in which the LC optic axis is oriented perpendicularly to the boundary surfaces which enclose the nematic layer. The uniform homeotropic orientation is obtained with some nematics without pretreatment of the boundary surfaces apart from a very careful cleaning. Other nematics require an additive such as hexadecyltrimethyl-ammonium bromide (Eastman Organic Chemicals No. P5650). For example, a 0.5% solution of the additive in MBBA (room temperature nematic LC, Eastman Organic Chemicals No. 11246) by mixing 0.0024 and 0.495 grams respectively in a vial, heating to about 50°C and shaking it until all the solid mass is dissolved. A drop of the solution at room temperature is placed between un-treated boundary surfaces and the nematic liquid crystal (NLC) film assumes a homeotropic (perpendicular) texture. Another method of obtaining this texture is to coat the boundary surfaces with a thin layer of a surfactant such as lecithin.

Thin Mylar or Teflon films are satisfactory as spacers. These non-conductive, inert gaskets keep the glass slides 1/4 - 2 mils apart (1 mil = 0.001 inch = 25 μm). A "U" shaped spacer is convenient since the open edge provided by the "U" shape allows the application of the liquid crystal along the upper edge of the opening with a medicine dropper. Capillary forces then pull the liquid between the glass surfaces. A little air space for thermal expansion of the liquid crystal should be provided. After being filled with fluid, the cell's perimeter should be sealed. The assembled cell may be edge-sealed with molten paraffin wax or epoxy resin. A rapidly curing two-part epoxy system ("5-minute adhesive") is best because it sets rapidly but it is fairly soft in cured condition, so that strains do not build up.

Application of silver conducting epoxy along the 5-10 mm offset area (see Fig. 125) will help to insure uniform distribution of current between the transparent electrodes.

Some of the solvents that may be sued to dissolve liquid crystals are: Methanol, chloroform, ethyl acetate, tetrahydrofuran, 1,2-dichloroethane, 1,1,2-trichloroethane and dichloromethane.

Suggestions for further reading:
(Physical properties and applications of liquid crystals.
Tutorial and review articles).

J.L. Fergason, Molecular Crystals, vl p293 (1966)
W.H. Harper, Molecular Crystals, vl p325 (1966)
I.G. Chistyakov, Sov Phys - Uspekhi v9 p551 (1967)
Groupe des cristaux liquides d'Orsay, La Recherche v2 p433 (1971)
R. Steinstraesser and L. Phl, Angew Chem Int'l Edit v12 p617 (1973)
G.H. Brown, J Opt Soc Am v63 p1505 (1973)
V.I. Lebedev, V.I. Mordasov and M.G. Tomilin, Sov J Opt Technol v41 p403 (1974)
L.M. Blinov, Sov Phys - Uspekhi v17 p658 (1975)

" " " " " " " " " " " "

Nematic Liquid Crystal Polarization Rotator and Depolarizer

Conventionally, half-wave retardation plates are used to rotate the plane of polarization through an arbitrary angle. The use of a half-wave plate is limited to the wavelength for which it was manufactured.

A nematic liquid crystal (NLC) film between two surface oriented (rubbed) glass plates becomes a device for rotating the plane of polarized light. If the two glass surfaces are superposed so that the direction of rubbing is parallel for the two surfaces on either side of the NLC film, the film behaves as a uniaxial solid with the optic axis lying in the direction the glass plates were rubbed. Linearly polarized light entering with its polarization direction either parallel or perpendicular to the optic axis of the film exits with the polarization direction unchanged. If the rubbing directions of both glass plates lie at an angle of 45-degrees with respect to the plane of polarization of the entering light, the exiting light will be depolarized.

Next, the two glass plates are arranged with their rubbing directions at 90-degrees to each other. The entering polarized light is again parallel or perpendicular to the orientation of the first rubbed surface. The direction of the plane of polarization of the exiting light will be aligned in accordance with the orientation direction of the second rubbed surface: it will be rotated through an angle of 90-degrees. Thus, by rotating the second surface, or by changing the rubbing direction of the second surface, the plane of polarization of the exiting light can be rotated at will to any desired angle relative to the plane of polarization of the entering light.

As the rubbed glass plates are rotated relative to each other, a helical twist is superposed on the nematic structure. The device is wavelength-independent since the angle of polarization rotation depends only on the angle of twist in the nematic structure.

Experimental cells can be prepared by placing a thin "U" shaped Mylar spacer between two pieces of carefully cleaned Kodak Slide Cover Glass plates, 50 x 50 mm size. Either dry rubbing or wet calcium carbonate, or a water slurry of rouge and a felt bar may be used for the rubbing operation. After rubbing the plates about two dozen times each, on one side, they are carefully rinsed in distilled water, rinsed in methanol, and dried without touching the rubbed surfaces. The NLC is applied with a medicine dropper to the open edge of the sandwich cell. Licristal #5 or #9 (manufactured by E. Merck, distributed in the USA by E.M. Labs) can be used as NLC for the polarization rotator and depolarizer device.

References:

I. Haller and M.J. Freiser, IBM Tech Discl Bull v13 p1211 (Oct '70)
J.F. Dreyer, U.S. Pat. No. 3,592,526 (13 Jly '71)

Distortion of Aligned Phases (DAP-Effect)

This is an electro-optical effect. Nematic liquid crystals of negative dielectric anisotropy are used. The initial orientation of the NLC-molecules is perpendicular (homeotropic) to the electrodes. Such layers behave like optically positive, uniaxial crystals. If such a NLC layer is placed between crossed polarizers the system cannot transmit light. If an increasing potential is applied to the electrodes, the molecules become increasingly parallel to the electrode surfaces. The result is that birefringence occurs and some light passes through the second polarizer (Figs. 126-a and 126-b).

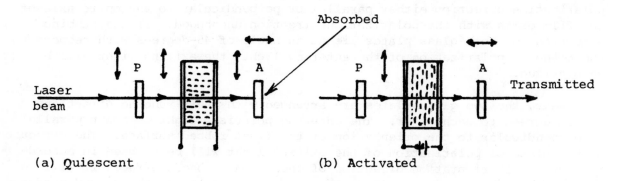

(a) Quiescent   (b) Activated

Fig. 126. Demonstration of the DAP-effect.

Current flow is not necessary for operation. NLC cells of this type are called field-effect devices. High-purity, high-resistance NLCs are used, such as Licristal #7B or #9B (E. Merck). The DAP-effect begins at a defined threshold potential of about 5-10 volts. When the potential is switched off, the molecules return to the homeotropic starting orientation and the system returns from transparency to opacity.

The Freedericksz-effect is the counterpart of the DAP-effect. The dielectric anisotropy required is positive (Licristal #10 or #684). The starting alignment of the molecules is parallel (homogeneous) to the conductive surfaces and the layer shows birefringence. The electric field realigns the molecules homeotropically and the birefringence vanishes. Such a layer between crossed polarizers can be switched from transparency to opacity, i.e., from light to dark.

One can apply a D.C., sinusoidal, or pulsed voltage to achieve these effects.

Reference:

Freedericksz, V.K. and V. Zdina, Trans Faraday Soc v29 p919 (1933)
M.F. Schiekel and K. Fahrenschon, Appl Phys Lett v19 p391 (1971)
G. Assouline, M. Hareng and E. Leiba, Electron Lett v7 p699 (1971)

The Twisted Nematic Effect

This electro-optical effect in twisted nematic liquid crystals was discovered by Schadt and Helfrich. Twisted NLC cells produce high contrast, exhibit low threshold potentials, driving voltages of 2-5 volts, and power consumptions of 1-50 $\mu W/cm^2$.

The thin electro-optical cell consists of two glass plates coated on the interior surfaces with a transparent conductive material (NESA or NESATRON Glasses). The interior surfaces of the cell comprising the electrodes are rubbed uniformly along one direction. When the plates are brought together to form the cell, they are positioned so that the rubbing directions are at right angles to each other. The two plates are spaced apart by suitable spacers. The space between the plates is filled with a NLC material with positive dielectric anisotropy, such as Licristal #10 or #684 (E. Marck). In the absence of an electric field, the liquid crystals align with their molecular axes parallel to the surface of the plates (homogeneous alignment) and along the rubbing direction. With rubbing directions at right angles to each other, the molecular axes are oriented so as to form a helix in the cell and polarized light transmitted through the cell is rotated 90 degrees. When such a cell is placed between two parallel polarizers, no light will be transmitted at zero voltage. If an electric field is applied to the electrodes, the structure will untwist at a well defined voltage and the optical activity vanishes thus allowing light transmission. The opposite effect, switching from light to dark, can be brought about if the cell is placed between crossed polarizers.

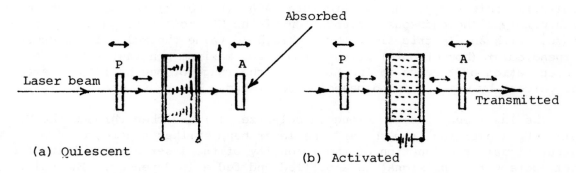

Fig. 127. Twisted nematic cell between parallel
polarizers.

By applying D.C. or A.C. voltages the rotatory power of the cell can practically be reduced to zero since the effect of the electric field is to induce homeotropic orientation of the molecules. The effect shows no hysteresis and the cell returns to its original homogeneous state when the voltage is turned off. Under D.C. operation higher voltages are needed for the effect to occur. The preferred mode of operation is with A.C. potential from 20 Hz to 1 kHz.

Reference:

M. Schadt and W. Helfrich, Appl Phys Lett v18 p127 (15 Feb '71)

The "Guest-Host" Effect

In 1968, G.H. Heilmeier and L.A. Zanoni reported a new electro-optic effect based on "guest-host" interactions in nematic liquid crystals (NLCs). The "guest-host" effect is observed when an electric field acts on a NLC (the host) containing as the additive (the guest) dichroic dye molecules that absorb light anisotropically. The orientation of the guest is controlled by the orientation of its nematic host, provided the additive molecule has an elongated shape.

Figure 128 is a schematic representation of a device for modulating a laser beam.

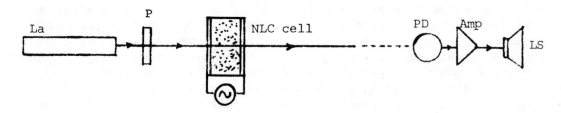

Fig. 128. Liquid crystal light modulator.

The liquid crystal cell (NLC) is the usual parallel plate capacitor (sandwich cell) with transparent electrodes (NESATRON Glass). The nematic liquid crystal plus dye serves as the dielectric. Typical electrode spacings are approximately 12 μm. Room temperature nematic liquid crystal of positive dielectric anisotropy, such as Licristal #684 (E. Merck) is used as the host, with parallel (homogeneous) orientation of the NLC molecules in the quiescent state. With an electric field above threshold value the molecules assume perpendicular (homeotropic) orientation. For maximum contrast ratio the concentration of the dye in the NLC is in the range of one-half to 2 percent. One percent by weight of Indopehnol Blue or Sudan Black dye works well.

The laser beam passes through a polarizer (P) and then through the NLC cell. Upon emerging from the NLC the laser beam strikes a detector (PD). The output signal is a function of the intensity of the laser beam striking the photo-detector. The signal is amplified and fed a loudspeaker. The polarizer is oriented so that a maximum amount of light is absorbed when there is no field applied to the cell. If A.C. is used, it is possible to obtain modula-tion of the laser beam by variation of the frequency as well as the voltage. There is no change in the absorbance with a small initial field. At fields above the threshold value there is a large change in absorbance.

References:

G.H. Heilmeier and L.A. Zanoni, Appl Phys Lett v13 p91 (1 Aug '68)
J.A. Castellano, U.S. Pat. No. 3,597,044 (3 Aug '71)

Liquid Crystal Temperature Detection Device

Optical activity in mesomorphic media was first observed in derivatives of cholesterol, and media which show this phenomenon are therefore termed cholesteric media.

Cholesteric liquid crystals (CLCs) possess anomalously high optical activity which changes sign at some critical wavelength λ_o where $\lambda_o = 2np$, with n representing the average index of refraction of the CLC (about 1.5), and p the pitch or repetition distance of the helical structure of the liquid crystal material. Thin films (0.5 to 50 μm) of CLC substances rotate the plane of polarization by substantial amounts depending upon both the wavelength of the incident light and the pitch of the liquid crystalline substance. Rotations of the order of 50,000 degrees per millimeter may be observed in the mesomorphic state, but if the temperature is raised until the anisotropic liquid becomes isotropic, the rotatory power falls at once to the almost immeasurably small rotation per millimeter which is associated with ordinary optically-active liquids and solutions.

The pitch of these CLC substances can be varied by the presence of some stimulus to which their pitch is sensitive, e.g., temperature, pressure, shear, electric and magnetic fields, etc. Thus, varying the pitch in response to some stimulus results in a change in the amount of rotation imparted to the incident light by the liquid crystal film.

Optical activity is observed by illuminating the CLC film with linearly polarized light. Due to the optical activity of the material (also called optical rotatory power), the polarization vector of the light is caused to rotate. The amount of this rotation has been found to be dependent upon the temperature of the material.

Figure 129 shows a temperature detection apparatus, comprising a laser (La), a linear polarizer (P), a cholesteric liquid crystalline film (CLC), a linear polarization analyzer (A), a photo-detector (PD), an amplifier (Amp), and a galvanometer (M).

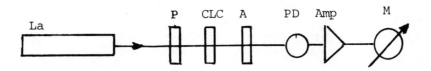

Fig. 129. Liquid crystal temperature
detection device.

A thin film of a cholesteryl chloride (35% by weight) and cholesteryl nonanoate (65% by weight) mixture is formed on a glass slide and another glass slide is placed over the free surface of the liquid crystal film. The beam of a He-Ne laser having a wavelength of 633 nm is directed upon the CLC film at normal incidence and at room temperature. The linear polarization analyzer is arranged behind the CLC film with its axis of polarization adjusted so that the incident light is completely extinguished.

The CLC film is then heated with a heat gun at gentle heat for a few seconds. A relatively small temperature change will produce a relatively large change in the amount of rotation imparted to the incident light and some light will be observed emerging from the linear analyzer. The D.C. voltage output of the photo-detector is a measure of temperature.

"Licristal' (E. Merck) temperature indicators can be used to prepare a liquid crystal layer which shows iridescent colors at a temperature slightly above room temperature. Either Thermomagic (TM) color 28/30°C or 30/35°C preparations may be used. They are available in the pure state, without added solvents. These pastes are readily soluble in trichloroethylene, carbon tetrachloride, chloroform, benzene and other solvents for fats and fatty materials.

The liquid solution is applied on a small piece of conducting glass (NESA Glass) and the solvent is allowed to evaporate, resulting in a thin layer of CLC. The CLC film is sealed from the atmosphere with a microscope slide. The CLC film is then heated by sending a small, adjustable current through the conducting glass.

References:

L. Melamed and D. Rubin, Appl Opt v10 p1103 (May '71)
L.B. Leder and D. Olechna, Opt Commun v3 p295 (Jly '71)
H.J. Gerritsen and R.T. Yamaguchi, Am J Phys v39 p920 (Aug '71)
J.E. Adams and W.E.L. Haas, U.S. Pat. No. 3,726,584 (10 Apr '73)
J.E. Adams and W.E.L. Haas, U.S. Pat. No. 3,848,965 (19 Nov '74)

" " " " " " " " " " " .

Interference and Interferometers

If light from a source is divided by suitable apparatus into two or more beams which are then superposed, the intensity in the region of superposition is found to vary from point to point. There will be maxima, which exceed the sum of the intensities in the beams, and minima which may be zero. This phenomenon is called optical interference. Interference cannot occur unless the beams are coherent with respect to each other. In the laser we have a very bright source of extremely coherent light.

The basic system of an interferometer is shown in Figure 130.

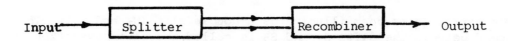

Fig. 130. Basic system of an interferometer.

There are two general methods for obtaining beams from a single beam of light: (1) The beams are divided by passage through apertures placed side by side (Fig. 131). This method is called "division of wavefront." The two beams are then superposed in the region in which the interference phenomena are observed. Young's slits, Fresnel's biprism and Lloyd's single mirror are based on this principle. (2) The beam is divided at one or more partially reflecting surfaces, at each of which part of the light is reflected and part transmitted (Fig. 132). This method is called "division of amplitude." The two beams are then recombined to obtain interference patterns.

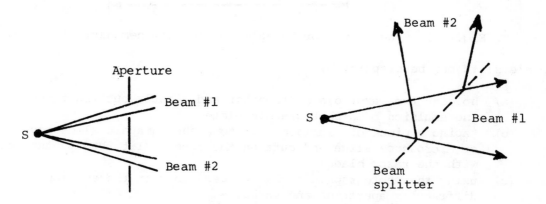

Fig. 131. Interference by division of wavefront.   Fig. 132. Interference by division of amplitude.

The interference produced by plane parallel plates and the formation of Newton's rings is based on this principle. A more modern use is in Michelson's interferometer and many other interferometric devices.

It is convenient to consider separately the effects which result from the superposition of two beams ("two-beam interference"), and those which result from the superposition of more than two beams ("multiple-beam interference").

Suggested Additional Readings:

J.D. Briers, Opt & Laser Tech p28 (Feb '72)
W.J. Smith, Opt Spectra p36 (Apr '74) and
W.J. Smith, Opt Spectra p52 (May '74)
J. Vrabel and E.B. Brown, Opt Engineering p124 (Mch/Apr '75)

"""""""""""

Young's Double Slit Method for Producing Interference

The narrow laser beam illuminates a double slit (S_1 and S_2). The distance d between the slits S_1 and S_2 is called the slit separation. A beam expander is necessary for the demonstration when the separation between the slits is wider than the diameter of the laser beam.

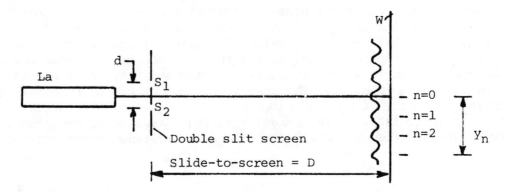

Fig. 133. The setup for Young's double slit demonstration.

Double slits may be prepared by

(a) holding two razor blades together and lightly cutting through the emulsion of a photographic plate.
(b) taping a piece of aluminum foil to a photographic slide cover or microscope slide and cutting the double slit through the foil with the razor blades.
(c) using the photographic technique described in the section on diffraction apertures and masks.

The double slit is placed in position on the optical bench to intercept the laser beam. The interference pattern can be observed on the distant screen, W.

The point O on the screen (Fig. 134) is a position of maximum constructive interference and appears brightly illuminated because O is equidistant from S_1 and S_2. The difference in path lengths (between r_1 and r_2) to the point P on the screen is approximately $e = d \sin \theta$. The wavelets from S_1 and S_2 reinforce each other at P if the path difference e is equal to a whole number n of wavelengths ($e = n\lambda$), that is, crests of both waves from S_1 and S_2

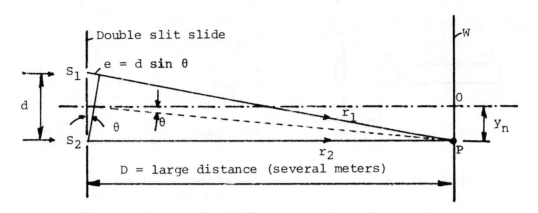

Fig. 134. Phase relation diagram.

arrive at P and there appears a bright light fringe. If the path difference $e = (n+1/2)\lambda$, a crest and a trough arrive, giving an intensity minimum.

The position of the maxima on the screen are at

$$y_n = D\theta = Dn\lambda/d$$

By measuring the separation d between the slits, the distance D between the double slit slide and the distant screen, and the distance y_n from the center of the zeroth fringe 0 to the center of the n-th fringe on either side, the wavelength of the light producing the interference pattern may be computed:

$$\lambda = y_n d/nD$$

Since the laser radiation wavelength (λ=633 nm for He-Ne lasers) is known, by measuring D and y_n we can calculate the slit separation d from

$$d = n\lambda D/y_n$$

The same d can be determined from the dark fringes by using the formula:

$$d = (n+1/2)\lambda D/y_n.$$

" " " " " " " " " " "

Interference with Fresnel's Biprism

The Fresnel biprism is the easiest device for obtaining the interference between two sections of a wavefront.

The biprism is usally made of a single piece of glass. It may be regarded as consisting of two equal triangular prisms base-to-base and with very small refracting angle α (>1°).

The laser beam is "cleaned" with a lens-pinhole spatial filter (SF). The diverging beam is made to cover the obtuse angle of the biprism. Fine fringes can be seen on a screen 2-3 meters away from the biprism. A converging lens

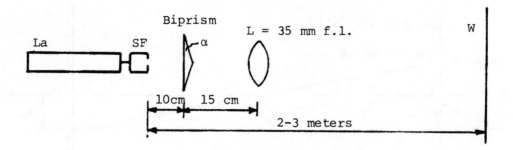

Fig. 135. Fresnel biprism demonstration setup.

(L) placed downstream projects an enlarged pattern on the screen. This setup
is suitable for lecture demonstrations of Fresnel biprism fringes.

To find the wavelength λ of the laser light, it is necessary to know the
distance d between the virtual images S_1 and S_2 from which the rays bent by
the biprism seem to come. The value of d is obtained experimentally as
shown in Fig. 136.

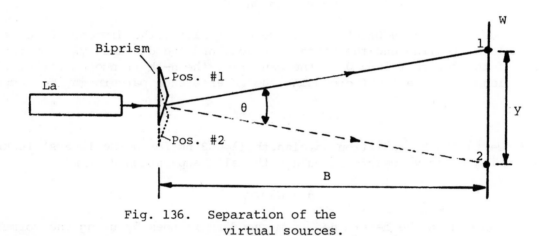

Fig. 136. Separation of the
virtual sources.

The direct laser beam is used without the lens-pinhole filter. In Pos.#1
the lower half of the biprism is effective and a bright red spot appears on
the screen. The biprism is then moved across the laser beam to Pos. #2 and
the spot now appears at "2" instead of "1" on the screen. In other words, the
laser beam first passes through the lower half and then through the upper half
of the biprism. The center of spots 1 and 2 are marked on the screen and the
linear displacement Y measured. Then $\theta = Y/B$, where B is the biprism to screen
distance.

Since θ is the angle subtended by S_1 and S_2 at the biprism, then $d = A\theta$.
The spacing Δy of the interference fringes formed on the screen is given by
the Young's double slit equation

$$\Delta y = D\lambda/d.$$

The fringe spacing Δy is determined by measuring the distance covered by
several dark or light fringes and dividing.

Writing the above equation in terms of directly measured quantities as

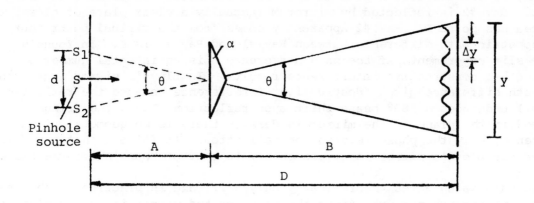

Fig. 137. The biprism exercise to measure the
wavelength of laser light.

shown in Fig. 137, and solving for λ, we have the formula

$$\lambda = \Delta y A Y / D B.$$

Measurement of the distance D can be done conveniently by using a long piece of string.

References:

F.Y. Yap, Am J Phys v37 p1204 (1968)
A.L. Shawlow, Am J Phys v33 p922 (1965)
J.J. Veit and D.J. Solarek, Phys Teach p413 (Oct '75)

" " " " " " " " " " "

Demonstration of Lloyd's Mirror Fringes

Lloyd's mirror is a classic method for obtaining interference fringes. First described in 1834, it is a simple technique for finding the wavelength of light. The geometry is the same as for Young's double slit interference.

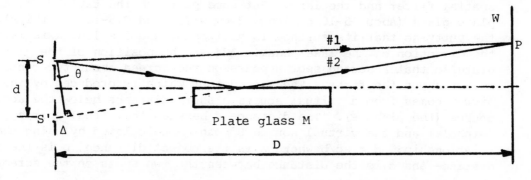

Fig. 138. Principle of Lloyd's mirror interference.

Ray #1 emerges from a pinhole and travels directly to point P on the screen. Ray #2 is reflected by mirror M (actually a clear plate of glass) and also reaches point P. Ray #2 apparently comes from the virtual point source S'. The geometric path difference between Ray #1 and Ray #2 is Δ. In Young's double slit experiment, if the path difference Δ is an integral number of wavelengths, constructive interference results. In Lloyd's mirror experiment this path difference gives "destructive" interference because the reflected ray (#2) undergoes a 180° phase shift upon reflection. Thus, the fringe observed in the plane of the mirror is dark, - there is no geometrical path difference, only the phase reversal on reflection. It follows that the fringe pattern for Lloyd's mirror is complementary to that of Young's interferometer.

Lloyd's setup gives theoretically only half, and in practice, much less of the interference picture, since the plane of the mirror is in the plane of symmetry, and the lower half of the pattern is obstructed by the presence of the mirror.

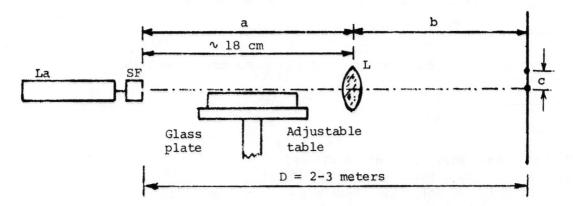

Fig. 139. Lloyd's mirror setup (not to scale).

Figure 139 shows the setup for the Lloyd's mirror exercise.

1. Attach the laser to the adjustable laser carrier and place the carrier on the optical bench. Screw the lens-pinhole spatial filter into the laser housing and adjust it to obtain a "clean" patch of red light on the screen about two meters away. Mount the converging lens (L = +167 mm focal length), on the optical bench. Adjust the lens so that the laser beam produces the smallest possible spot on the screen.
2. Place a component carrier with an adjustable table between the spatial filter and the lens. Put some putty on the table. Lay a good plate glass (about 5-10 cm long, 2 cm wide, and 0.5-1.0 cm thick) on the putty so that its surface is horizontal and its long axis is in the direction of the laser beam. Adjust the position of the glass plate so that a second spot appears on the screen about two centi- meters above the first. The second spot which is produced by the mirror comes from a virtual source a short distance below the actual source (the pinhole). The distance d between the actual source S (pinhole) and the virtual source S' can be calculated by using the relationship: d = ac/b where a is the object distance, b is the image distance and c is the distance between the two spots on the screen.

3. Carefully remove the converging lens L from the optical bench without disturbing any of the other components. Without this lens, the cones of light coming from the two sources (S, S') will partially overlap forming an interference pattern of alternate red and black horizontal bands or fringes along the bottom of the illuminated area on the screen. Once the fringes are obtained, a more delicate adjustment of the fringe spacing can be obtained with the leveling screws on the adjustable laser carrier. Note: Even when the two beams (direct and reflected) are separated, two sets of fringes are observed. These are caused by diffraction effects at the edges of the mirror. After the two beams are made to overlap by raising the mirror, the interference fringes are definitely more distinct.

4. The wavelength of the laser light can be calculated by using the formula given for double slit interference:

$$\lambda = d \sin \theta$$

where: λ is the wavelength of the laser light, d is the distance between sources S and S' as determined in step #2, and $\sin \theta$ is the distance between any two bright or any two dark fringes on the screen.

Note: The distance between fringes on the screen is easily measured if a graph paper is used as the screen. If there is any difficulty in counting the fine fringes, a magnifying glass held near the screen will be useful.

" " " " " " " " " "

Interference Demonstration with Plane Parallel Glass Plates

The purpose of this exercise is to demonstrate the interference phenomenon for light reflected from the boundaries of a thin parallel-sided glass plate.

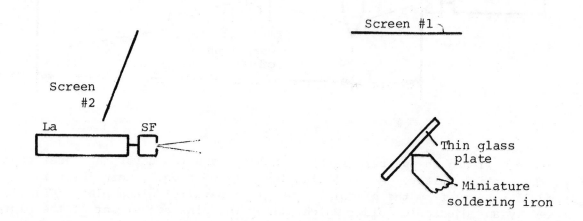

Fig. 140

Figure 140 shows one method of setting up such demonstrations The diverging light from the lens-pinhole spatial filter falls on a thin plane-parallel glass plate situated 25-50 cm from the spatial filter, at an angle of 45-degrees. The interference fringes are seen on Screen #1. By turning the glass plate so that its surface is almost perpendicular to the axis of the light cone incident on it, the fringes can be observed on Screen #2. Poor fringe regularity indicates variation in the optical thickness of the slide. Excellent fringes were obtained with Kodak Slide Cover Glass (B351, 50x50 mm, thin). Fringe spacing is larger on Screen #2 than on Screen #1.

If a warm soldering iron is held for a few seconds against the far side of the glass plate the latter is heated locally at the point of contact. The slight expansion which takes place is sufficient to cause a slight displacement of the fringes around the spot heated. The result is shown in Fig. 141. The fringes are displaced locally toward the thinner part of the glass plate.

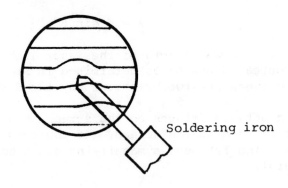

Fig. 141

Figure 142 shows a second method of obtaining the interference pattern with a plate glass.

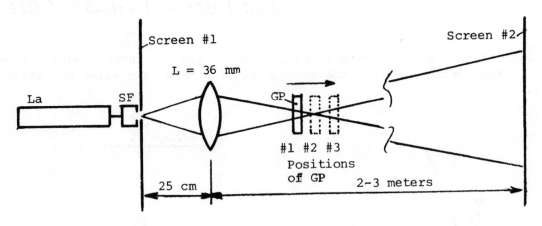

Fig. 142

Screen #1 has an aperture large enough to pass the diverging laser beam near the spatial filter. A converging lens (L) of focal length 36 mm is placed 25 cm away from the pinhole. The plane-parallel glass plate (GP) is carried by an adjustable holder which allows the fine adjustment of the plate to set it perpendicular to the axis of the light cone incident on it.
A system of interference rings is observed on Screen #1 in reflected light

and on Screen #2 in transmitted light. Repeat the exercise in converging beam (Pos. #1), at the focal point of L (Pos. #2), and in diverging light (Pos. #3). Observe the size and quality of the interference rings in the various positions and on both screens. Try the exercise with plane-parallel glass plates from 0.2 mm to 15 mm thickness.

A somewhat simplified version of the previous exercise is shown in Fig. 143.

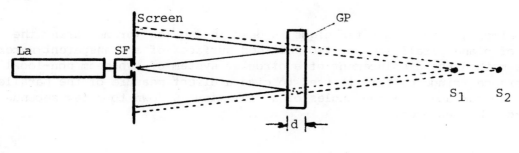

Fig. 143

The glass plate (GP) is set perpendicular to the axis of the light cone originating at the pinhole (S) of the spatial filter. (A good inexpensive glass plate fro this exercise is Edmund Cat. No. 30451, Flat Glass, 37 mm dia. thickness: 6 mm). Consider two interfering rays produced by reflection from the front and rear surfaces of the plate. These rays may be conceived as coming from two virtual sources, S_1 and S2, of the point source S (the pinhole). The distance between S_1 and S_2 is equal to 2d/n, where d is the thickness of the plate and n its refractive index. The interference fringes (rings) can be observed on the screen surrounding the pinhole.

Another simple setup is shown in Fig. 144. This arrangement permits us to view interference effects in reflected and transmitted light side-by-side.

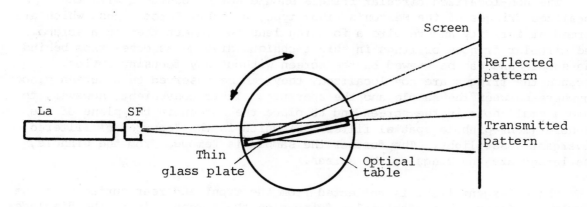

Fig. 144

The filtered and slightly divergent laser beam is incident on a thin glass plate (slide cover glass or microscope slide) which is attached to the rotatable optical table with a small amount of putty. The optical table is placed about 10-20 cm from the spatial filter and the screen about 2-3 meters further away. The optical table is turned until the glass plate is almost parallel to the beam. Two sets of interference patterns will be produced on the screen. Observe the two patterns and note the similarities and differences between the two patterns.

" " " " " " " " "

Interferometer for Measuring Parallelism and Small Wedge Angles

A simple, easy to use technique is described below for measuring the degree of plane parallelism of two polished surfaces of a transparent material. In this method the displacement of a circular system of two-beam non-localized interference fringe pattern is used to give a direct measure of the parallelism of two surfaces. Wedge angles of a few minutes down to a few seconds of arc can be measured.

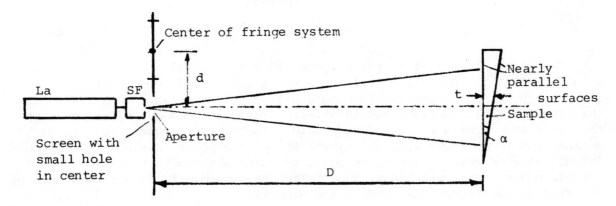

Fig. 145. Measuring parallelism and wedge angles.

The non-localized circular fringes should not be confused with the localized fringes of the Newton's rings type, or Fabry-Perot rings, which are formed at infinity and require a focusing lens to project them on a screen. The circular fringes or rings in this technique diverge as cones from behind the sample and can be viewed on the screen without any focusing optics. Because the fringes are non-localized, they can be observed on a screen placed anywhere between the sample and the aperture. It is convenient, however, to have a small hole in the screen and to place the screen in the plane of the pinhole (lens-pinhole spatial filter: SF) which is the "source" of filtered, divergent laser light. The further the sample is removed from the pinhole, the larger are the rings on the screen.

The divergent light is reflected from the front and rear surfaces of the sample and forms a set of circular fringes on the screen. It is the displacement of the center of this fringe system from the center of the screen that is a measure of nonparallelism of the two surfaces. The sample can be hand-held, and its positioning and rotation by hand are all that is required to locate the center of the fringe system on the screen. The center of the fringe pattern is always in line with the thick side of the sample and, thus, one can readily determine which part needs to be removed to achieve the desired parallelism.

For a given sample wedge angle and for a given screen-to-sample distance, the center of the fringe system occurs at a fixed distance from the center of the screen. If the wedge angle is zero (i.e. the surfaces are perfectly plane parallel), the center of the fringe system occurs at the center hole of the screen (aperture) and coincides with the pinhole through which passes the input divergent laser beam.

The wedge angle is given by the relationship

$$\alpha = (td)/(2n^2D^2) \text{ (radians)}$$

where t is the average thickness of the sample, n is the refractive index, D is the distance between the screen and the sample, and d is the displacement on the screen of the center of the fringe system from the center of the aperture. The wedge angle (α) is therefore directly proportional to d and can be determined from a single measurement of d, when t, n, and D are known.

References:

J.H. Wasilik, T.V. Blomquist and C.S. Willett, Appl Opt v10 p2107 (Sep '71)
C.S. Willett, Opt Spectra p36   (Dec '71)
T.G. Bergman and J.L. Thompson, Appl Opt v7 p923 (May '68)

""""""""""

Interference Demonstration with a Small Glass Tube

A He-Ne laser is adjusted to be incident at a near grazing angle on the surface of a vertical glass cylinder 1.0 cm in diameter and 1.0 mm wall thickness.

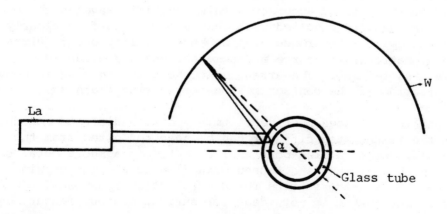

Fig. 146. Arrangement for producing interference fringes.

If a 180-degree screen, W, is placed around the cylinder at a distance of about 2.0 meters from the cylinder center, an interference pattern will be

observed consisting of vertical light and dark bands. The prominent fringe
pattern is due to interference of two singly-reflected rays: the ray reflected
from the outer surface of the cylinder and a ray reflected from the adjacent
inner surface. A laser beam having lateral coherence is necessary for fringe
formation because the two singly-reflected rays that interfere come from
different parts of the incident beam's cross-section. The fringe spacing
increases as the angle α becomes small. Increasing the wall thickness
decreases the fringe spacing.

References:

D.C. Brown and T.L. Rome, Am J Phys v40 p470 (Mch '72)
W.H. Southwell, Am J Phys v41 p284 (Feb '73)
B.B. Loud, P.L. Sardesai and S.H. Behere, Am J Phys v41 p720 (May '73)
W.C. Maddox, B.W. Koehn, F.H. Stout, D.A. Ball and R.L. Chaplin,
                                    Am J Phys v44 p387 (Apr '76)

" " " " " " " " " "

Interferometric Flatness Testing

Fizeau bands or fringes result from essentially two-beam interference
occurring near a thin dielectric film. One must focus the eye on (or very
near) the film to see these Fizeau fringes: they are said to be "localized
in the film." When monochromatic light is used to observe the Fizeau
fringes, they are interpreted in the same manner as the contour lines of a
topographical map. Fizeau fringes are called "fringes of equal thickness"
because the fringes, light and dark, are lines of constant separation between the surfaces and the spacing difference is one-half wavelength of the
illuminating light. An air wedge between two glass plates and Newton's
rings are each manifestations of Fizeau fringes.

Optical reference flats can be used to measure the flatness of machine
parts or of similar optical flats or mirrors. Each flat is interferometrically controlled in manufacture and furnished with a certification of
calibration as to its optical accuracy within 1/5, 1/10 wavelength, or
better. Flats are usually supplied with the reference surface clearly marked.
In addition to an optical reference flat of known quality one requires a
diffused monochromatic light source and some solvent and cleaning material
to perform flatness testing. The arrangement shown in Fig. 147 is convenient
for the demonstration of the contact method for testing flatness.

The surface to be tested must be illuminated with diffused monochromatic
light so that the illuminated diffuser can be seen reflected from the test
surface. The diffuser is a ping-pong ball cemented to a short metal or
plastic tube, which in turn, is attached with tape to the output window of
the laser. While this setup is fine for demonstration purposes, the
obliquity factor in viewing is not ideal for surface contour measurements.

Vertical illumination is extremely important and so is viewing at normal
incidence. The arrangement shown in Fig. 148 is highly desirable although a
more powerful laser is required because of the beam splitter.

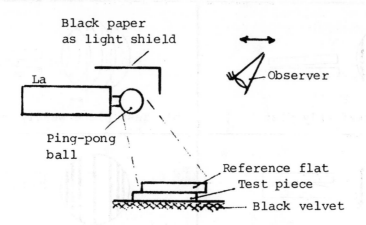

Fig. 147.   Illumination scheme for test-glass viewing.

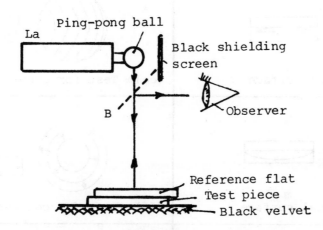

Fig. 148.   Beam splitter scheme to allow perpendicular
(normal) illumination and viewing.

Prior to actually contacting the reference and test surfaces, they must
be cleaned by the use of appropriate solvents and with an anti-static brush
to remove dust particles which remain after cleaning. The test piece is
placed on black velvet with the test surface uppermost and centrally located
under the ping-pong ball. A piece of lens tissue is now placed on the test
surface and the reference flat is laid on one edge. As the flat is gradually
lowered, the tissue is removed. When the tissue is completely removed, a thin
air wedge remains between the two surfaces. The two pieces are contacted with
light finger pressure until the number of fringes observed is a minimum. The
flat now sits on the "peaks of the hills" and the resulting fringes indicate
the "valleys" present in the test piece. Once fringes have been obtained it
is possible to rapidly assess the surface flatness of the component under test.

Figure 149 shows a few typical interference patterns which often occur
in practice (*).

(a)   Air-wedge - flatness error of zero: The fringes are straight,
      parallel and equally spaced.

(b)   Air-wedge - cylindrical error: In the example the overall wedge is

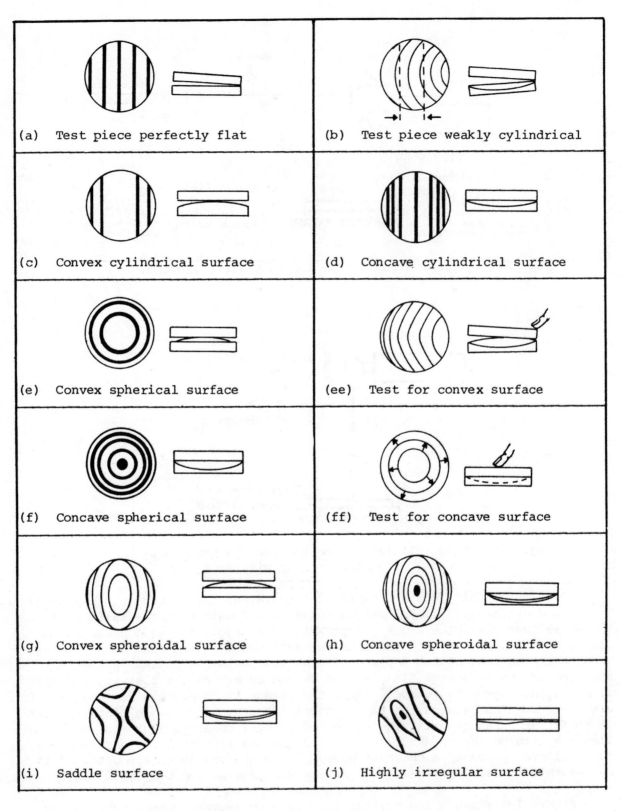

Fig. 149. Typical interference patterns and their interpretation (*)

6 fringes and the departure from flatness 2 fringes. When the surface is convex the fringes curve around the thin part of the wedge.

(c) Contact method - convex cylinder: There is a bright area in the center and a symmetrical fringe pattern with unequal spacing between successive fringes.

(d) Contact method - concave cylinder: A dark fringe exists in the center and the distance between successive fringes is decreasing as the periphery is approached.

(e) Contact method - convex sphere: A convex surface displays a bright patch in the center of the pattern. A test flat will "roll" on a convex surface if one edge is subjected to light pressure. The so-called "bull's eye" will move in the direction of the edge where the pressure is applied (see "e" & "ee").

(f) Contact method - concave sphere: A concave surface displays a dark fringe in the center of the pattern. If light pressure is applied at the center and the fringes move outward and reduce in number then the surface under test is concave (see "f" & "ff").

(g) Contact method - convex spheroid: A surface very close to spherical but slightly elliptical in shape is called a spheroid. The fringes are slightly elliptical with bright center.

(h) Contact method - concave spheroid: Elliptical array of fringes with a dark center.

(i) Contact method - saddle: A saddle type interference pattern is indicative of a surface possessing both positive and negative curvature about the central point in mutually perpendicular directions.

(j) Contact method - highly irregular surfaces: These matterns are found on low-quality; thin, plate glass.

References:

W.J. Smith, Opt Spectra, p36 (Apr '74) and p52 (May '74)

Material marked (*) reprinted with the permission of Edmund Scientific Co., from "Optical Techniques for Measuring Flatness" by Fred Abbott, copyright, 1974.

" " " " " " " " " "

Newton's Rings Demonstration

Newton's rings provide the classic illustration of thin film interference of the Fizeau type.

If a shallow convex spherical surface is placed on an optical flat, a system of circular fringes is seen around the point of contact. These are Newton's rings. The fringes (light or dark) are lines of constant separation between the surfaces. The spacing difference between two adjacent fringes is one-half wavelength of the illuminating light. Thus, the fringe pattern seen can be considered as a contour map of the space between the two surfaces. The ring system is also seen with transmitted light but with a much smaller contrast. These rings are exactly complementary to the reflected ring system.

Interference rings quite similar to "Newton's rings" can be demonstrated easily with a low power Ne-Ne laser.

When the traditional Newton's rings apparatus (long focus convex lens and a plane glass surface) is illuminated by laser light the resulting interference pattern will be complex. This is caused by the high coherence of the laser light; interference will take place between light reflected from the two surfaces of the lens and the two surfaces of the glass flat. When laser is used almost any thin lens will do and the optical flat of the conventional arrangement is not used. In our demonstration we use only the two surfaces of the thin lens for division of amplitude. The optical path difference at thickness t is 2nt, where t is the thickness of the lens and n is the refractive index of the glass.

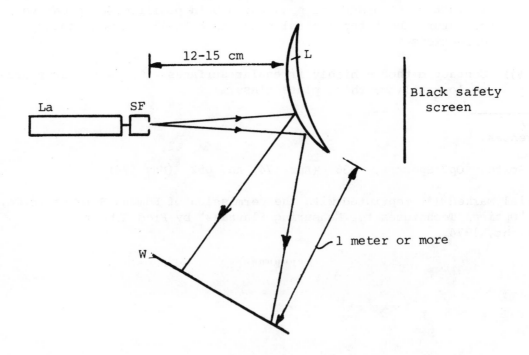

Fig. 150. Newton's rings demonstration setup.

Figure 150 shows the setup for Newton's rings demonstration. The raw
laser beam is cleaned and made divergent with the spatial filter (SF). A
long forcus lens (L) is placed in the beam about 12-15 cm from the pinhole.
Further downstream a black safety screen is placed to block the transmitted
light. A white cardboard is located near the laser about 1 meter away from
the thin lens. The convex side of the lens should face the cardboard (or
translucent) viewing screen (W). Two patches of red light will be produced on
the screen by reflections from the first and the second surface of the lens.
Careful horizontal, vertical and lateral adjustment of the lens will make the
patches overlap on the screen and a beautiful, contrasty ring pattern will
appear. Inexpensive, positive meniscus lenses of long focal length will
produce large ring patterns on a nearby screen. Fine results were obtained
with a 1,000 mm focal length positive meniscus (Edmund No. 94645) and also
with a 500 mm f.l. positive meniscus (Edmund No. 94646).

Reference:

M.J. Moloney, Am J Phys v42 p411 (May '74)

Fig. 151. Ring pattern produced with demonstration setup.

The collimation tester is a simple tool for adjusting laser collimator focus and for testing the planeness of a wavefront interferometrically.

Interference fringes are formed upon reflection of a laser beam from the front and back surfaces of a narrow wedge plate of glass. The same phenomenon can be observed with a thin film of air contained between two plane glass plates at a small angle to each other.

Let us assume that an absolutely plane wavefront is incident on the wedge with the narrow edge (apex) of the wedge either up or down. No matter in what direction, up, down or sideways the beam is deflected from the wedge, a screen intercepting the reflected beams will show horizontal fringes.

If the wavefront departs from being plane, for example, from a slightly defocused collimating lens, the fringe spacing changes if the beams are reflected up or down. When the beams are reflected to the right or left, there is a large change in the angle of the fringes to the horizontal.

Figure 152 shows the setup for focusing and testing a laser collimator.

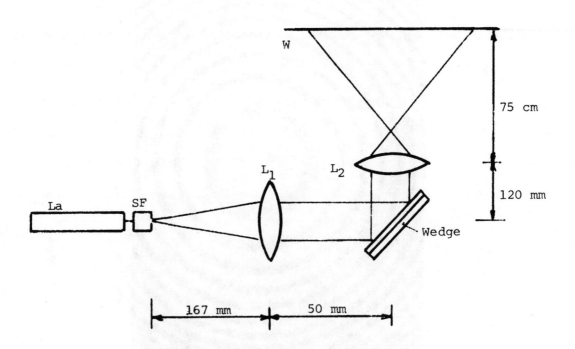

Fig. 152. Top view of collimation tester.

The laser beam is filtered and made divergent with the spatial filter (SF). The divergent beam is to be collimated by the lens (L_1 = 167 mm focal length convex lens). The lens is first adjusted for collimation by an auto-collimation method (described elsewhere in this book), using a plane mirror to return the light to the source (pinhole). Next, the mirror is removed from the optical path and the collimation testing device is placed in the collimated beam at a 45-degree angle to the axis of the beam. This angle is not critical. The thin edge of the wedge must be either up or down. Lens, L_2, approximately 36 mm f.l. is not absolutely necessary but it is useful in projecting an enlarged horizontal fringe pattern on the screen (W). If the collimated laser beam has a small diameter, the wedge is rotated from the 45-degree position to more normal incidence. The screen should then be moved to be normal to the reflected beam.

Two overlapping ellipses are seen on the screen. Since lens L_1 was pre-collimated, a fringe pattern should appear in the overlapping area. A slight movement of the collimator lens L_1 on the optical bench will introduce a large rotation of the fringe pattern. The lens is now carefully adjusted so that the fringes are horizontal again for best collimation. Curvature of the fringes or departure from straightness indicates errors or aberrations of the collimator lens which cannot be eliminated by focus adjustment.

References:

M.V.R.K. Murty, Appl Opt v3 p531 (Apr '64)
P. Langenbeck, Appl Opt v9 p2590 (Nov '70)
M. Lurie, Opt Engineering v15 p68 (Jan/Feb '76)

" " " " " " " " "

Thickness Measurement by Interference

If a transparent wedge of small angle, placed in air, is illuminated with laser light, an interference pattern is created. This pattern consists of a number of straight-line, equidistant fringes, called "lines of equal thickness."

To understand how this pattern is created, consider the light beam incident at an angle θ on a thin plate (or film) with refractive index n and a thickness t, as shown in Fig. 153.

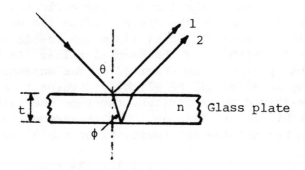

Fig. 153

Ray 1 is reflected at the top boundary while ray 2 is reflected on the lower boundary. When the two reflected parts of the incident beam recombine an interference pattern is generated. If the two rays are in phase maximum reinforcement occurs while, if the two are out of phase, the reinforcement diminishes. A minimum occurs when the phase difference is 180 degrees ($\lambda/2$).

The phase difference of the two rays is controlled by the difference in optical path length ($2tn \cos\phi$) and the phase shift of 180 degrees that occurs when the light reflects from a more dense to a less dense medium. A maximum light intensity occurs when the path length difference is $2nt = (m + 1/2)$, where m = 1, 2, 3,...

The above conditions still hold if the two boundary surfaces include a small angle such as a thin transparent wedge. The interference pattern created by a thin wedge consists of dark and bright fringes corresponding to an increase of the thickness of the wedge by $\Delta t = \lambda/2n$.

The separation of the fringes w is given by $w = \lambda/2n \, tg\alpha$, where α is the wedge angle.

As an exercise measure the thickness of a thin slide cover glass.

Set up the laser, the spatial filter, lens and glass plate as shown in Fig. 154.

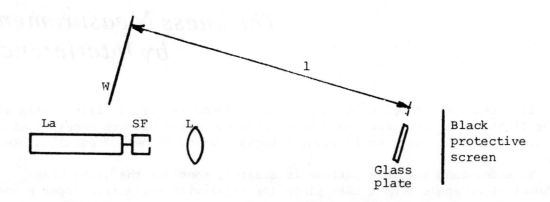

Fig. 154

Measure the angle between the incident and reflected beams at the glass plate. This equals 2θ. Measure the distance from the glass plate to the screen. This is the distance l. Initially place the lens L (50 mm f.l.) at 50 mm distance from the pinhole of the spatial filter. At this point two closely spaced light spots will be observed on the screen. Move the lens slowly away from the laser until the slightly divergent beams of light overlap and fringes are formed on the screen. Average the value of the separations between several bright or dark fringes. The measurement of the fringe separations, x, serves to measure the thickness, t, of the glass plate...

$$t \approx (nl\lambda)/(2x \sin\theta)$$

$\lambda = 6.328 \times 10^{-7}$ m, n = 1.5 for glass, l, x, and θ are measured values. Calculate the thickness of the glass plate from the above equation and compare the calculated value to what you measure directly with a scale.

Repeat the above exercise using a soap film instead of the glass plate. Place the wire frame for the soap film in a protective Lucite box. The wire frame should be aligned along the diagonal of the Lucite cube so that the reflected light comes out of the side of the cube as shown in Fig. 155.

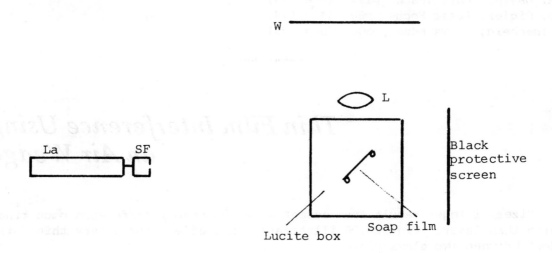

Fig. 155

The focal length of the lens, L, depends on the distance between the soap film and the viewing screen (W). Experiment with lenses of 36 mm to 200 mm focal length.

Once a soap film has formed, water will drain away from the film producing a wedge shaped cross section with constant horizontal thickness. Light striking the film will be reflected from both the outer surface and, after refraction, from the inner surface. Horizontal interference bands will be observed on the screen.

As the water continues to drain away from the film, the angle of the wedge shaped film decreases and the interference fringes move further apart. Eventually, there is only one fringe in view. When the soap film drains off to the point where the thickness of the film is much smaller than the wavelength of the light, the film becomes black. At this stage the phase difference is 180 degrees between the light reflected from the outer and inner surfaces of the film. The phase change of 180 degrees is due to reflection at the outer surface. Since the film thickness is much smaller than λ, the path difference between the two reflected rays can be neglected. Therefore, the two rays cancel to give zero intensity. The soap solution, called "Wonder Soap" (Chemtoy) has a drain-off time which is neither too slow nor too long, so that the effect can be conveniently observed.

Repeat the experiment with a piece of transparent plastic sheet. Excellent results can be obtained with "Stretch 'n Seal", a commercial food-wrapping material (Colgate-Palmolive Co.). This film is only 40 micrometers thick. It reflects about 8% of the incident light. The wrap material is manually stretched over a thin metal ring. It is secured to the sides of the ring with tape with two adhesive surfaces. The excess material is then trimmed off and the somewhat wrinkled film is tightened and stretched thinner by heating it for a few seconds with a hot-air blower. During heating the small blemishes and previously visible creases almost completely disappear.

_____

References:

J.A. Davis,   Phys Teach   p177   (Mch '74)
R.D. Sigler, Laser Focus p60   (Aug '74)
C. Isenberg, Phys Educ   p500   (Nov '75)

"""""""""""

Thin Film Interference Using
an Air Wedge

Fizeau fringes result from essentially two beam interference occurring near a thin layer. Figure 156 illustrates this effect for a very thin "air wedge" between two glass plates.

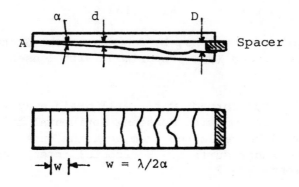

Fig. 156. The air wedge.

The upper plate is an optical flat and the lower another optical flat, or it may be a sample to be tested for flatness. If both pieces are very flat, a series of equidistant alternately bright and dark straight bands running parallel to the line of intersection (A) of the glass plates is observed. In the region of a slight depression in the lower plate, the gap thickness will increase and the fringes in the area will curve toward the wedge's apex (A).

Fizeau fringes are easily shown by using two pieces of plane parallel plate glass. They are laid together with a thin strip of paper along one edge as a spacer and then held together with variable pressure clamps at A and at the opposite edge where the spacer is located. Figure 157 shows setups for the observation of localized fringes of equal thickness, which are contours of the wedge thickness.

The interference fringes (that is, the loci of equal thickness of the air film) have a separation, $w = \lambda/2\alpha$.

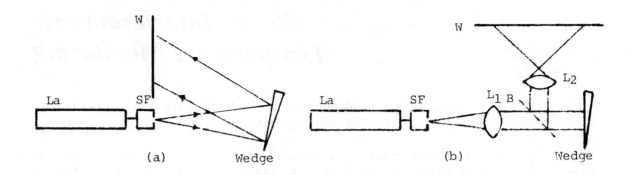

Fig. 157. Observing localized fringes of equal thickness.

The thickness of the film at the mth maximum is

$$d_m = (m + 1/2)(\lambda/2)$$

In going from one dark or bright fringe to the next, m increases by 1, and this requires that the optical thickness of the film, d, should change by $\lambda/2$ (for a near perpendicular observation direction). Using these Fizeau fringes one can measure the angle α of a very thin wedge=shaped sheet of any transparent material, or the wedge-shaped air gap between two slightly inclined plates.

Figure 157-a shows arrangement for viewing the fringe pattern.
Fig. 157-b depicts an alternative arrangement using a beam splitter. L_1 ia a 167 mm collimating lens and L_2 is a 36 mm projection lens. The screen is placed about 1-3 meters from the beam splitter.

_____

References:

Y.E. Amstislavskii, Sov Phys Uspekhi v10 p400 (Nov/Dec '67)
Z. Gubanski, Phys Teach p455 (Nov '69)

" " " " " " " " " "

Interferometric
Temperature Monitoring

A small He-Ne laser can be used to continuously measure the temperature of a transparent substrate during vacuum deposition.

In a vacuum deposition process, a metallic material is vaporized under intense heat inside a vacuum chamber and the metallic vapors are deposited onto a substrate. The laser beam is directed to the surface of the transparent substrate through the glass wall of the vacuum chamber. Some of the light is reflected from the surface of the substrate through the opposite wall of the vacuum chamber to a converging lens outside the chamber. The beam is received by a photo-detector whose amplified output drives a strip chart recorder. The schematic for the temperature monitoring apparatus using interference effect is shown below.

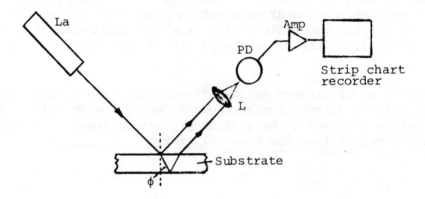

Fig. 158. Arrangement for temperature monitor.

Some of the laser light is reflected from the first surface of the substrate directly to the converging lens (L) while part of the light is refracted and then reflected from the rear surface through the converging lens to the photo-detector (PD). The intensity of the re-combined light beams monitored by the photo-detector will vary in accordance with the interference between the directly reflected beam and the refracted beam.

The thickness of the substrate material will vary with its temperature. As it expands or contracts, the conditions for light interference will change. The change in thickness can be detected by observing the fluctuations in light intensity as recorded on the strip chart.

The change in thickness of the substrate can be calculated by the expression $\Delta D = (m\lambda)/(2n \cos \phi)$, where m is the number of cycles recorded on the strip chart, λ is the wavelength of the laser light, ϕ is shown in the drawing, and n is the index of refraction of the transparent substrate. The change in temperature is then given by $\Delta T = \Delta D/K$, where K is the coefficient of linear expansion of the substrate.

With the change in temperature known via the above calculations, it is

only necessary to add this change in temperature to the original temperature
(room temperature) of the substrate to determine its present temperature.

References:

J.K. Alstad, V.Van Baren and J.R. Wiitala, IBM Tech Discl Bull
                                        v13 p2755 (Feb '71)
Lasersphere v2 p1 (Sep 15, '72)

"""""""""""

The Michelson Interferometer

The Michelson interferometer is an optical apparatus which includes a
beam splitter for dividing an input beam of light into two beams of approximately equal intensity. The two separated beams of light are made to traverse
paths of differing optical lengths and are subsequently combined. If the
optical path lengths do not differ by too large an amount, interference
effects occur.

The Michelson interferometer consists essentially of the above-mentioned
beam splitter (a semi-reflecting glass plate having parallel faces lightly
silvered on one surface), being positioned so as to be at 45 degrees to an
incident beam of light, and two plane front-surface mirrors situated at right
angles to each other and at 45 degrees to the semi-reflecting glass plate.
The beam splitter produces from the incident beam of light two beams having
approximately equal intensities, one being reflected from the front surface of
the beam splitter and the other beam being transmitted through the back surface of the beam splitter. Each beam of light is so reflected from one of the
plane mirrors that the beams retrace their paths. After reflection from the
plane mirrors, the two beams are recombined at the beam splitter to form an
output beam (at 90 degrees to the incident beam) capable of presenting an
interference pattern.

The interference pattern will vary cyclically with the relative displacement of the two mirrors in directions normal to their surfaces. One of
the mirrors usually remains fixed and the optical path from the light source
to the point of recombination of the split beams by way of the fixed mirror
is referred to as the "fixed leg" of the interferometer. The optical path
from the light source to the point of recombination of the split beams by way
of the movable mirror is referred to as the "variable leg" of the interferometer. A relative movement between the mirrors of half a wavelength of the
incident light will cause the intensity of the output beam to vary through
one cycle.

The Michelson interferometer has been used for measuring unknown wavelengths; to determine accurately lengths such as the international standards
of length (standard meter) and also to study the fine structure of spectrum
lines. Another application is to accurately monitor linear movement, the
movable mirror being attached to the body, the motion of which is monitored.
The interferometer is also capable of measuring lateral displacements, pitch,

yaw and roll of a moving body. Other applications of this versatile instrument include the determination of the index of refraction of a gas and of various transparent solids.

Most of the applications mentioned above are not suitable for laboratory experimentation. The original Michelson interferometer suffers from some disadvantages. It is very sensitive to misalignment of the mirrors because the tolerance of angular tilt in the motion of the movable mirror is only a few seconds of arc. As the interferometer is made up of three separate optical components, it will have eighteen degrees of freedom, fifteen of which affect the fringe pattern.

The interferometer to be described is similar in arrangement to the original Michelson interferometer. The laser interferometer differs from the classical setup by the different treatment of the test beam and the reference beam. This is possible because of the unique coherence properties of lasers. The interferometer can be set up and operated by persons not experienced in the use of high-precision optical instruments.

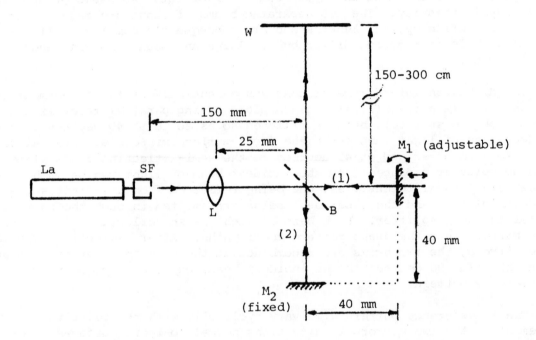

Fig. 159. Optical diagram of Michelson-type interferometer.
(Not drawn to scale)

The components of the Michelson-type interferometer are arranged as shown in Fig. 159. By following the procedure outlined below, the interferometer may be quickly aligned.

1. Mount the laser on the adjustable laser carrier and place it on the optical bench.

2. Affix the lens-pinhole spatial filter (Mtrologic No. 60-618) on the laser and adjust it to obtain a clean, slightly divergent beam.

3. Place the beam splitter assembly (Metrologic No. 00111) on the optical

bench. Approximate distance between the center of the beam splitter should be about 150 mm.

4. Position the carriage assembly (Metrologic No. 00112) with adjustable mirror M_1 on the optical bench close to the beam splitter assembly (Metrologic No. 00111). The distance between the adjustable mirror, M_1, and the center of the beam splitter should be 40 mm.

5. Set up the fixed mirror, M_2, 40 mm distance from the center of the beam splitter.

6. Place a screen (white cardboard), W, about 1.5 to 3.0 meters from the beam splitter to receive the recombined beams.

7. Adjust the laser so that the beam is centered on the adjustable mirror, M_1. The lens, L, is not used at present and it should not be in the beam path.

8. Position mirror M_2 so that the laser beam reflected by the beam splitter strikes its center. Two red circular light patches should now be visible on the viewing screen.

9. By turning the adjusting screws which control the orientation of M_1 (on the carriage assembly), the two light patches on the screen are brought into coincidence. A slight adjustment of the laser and the table carrying the beam splitter may also be required at this point. It may be necessary to bring the two patches past the position where they should be superposed before fringes are observed. After fringes have been obtained, the center of the fringe system can be brought into view by further small adjustments of the two micrometer screws. Note: The screws should be turned a little at a time and in such a direction as to increase the curvature of the fringes. The more precise the adjustment during the previous steps, the wider the fringes will be. Try to fill the screen with one fringe only.

10. To obtain an enlarged fringe system on the screen, introduce converging lens, L, in the beam path as shown in Fig. 159. This is a 36 mm focal length lens located approximately 25 mm distance from the center of the beam splitter. By adjusting L, the display on the screen can be optimized.

11. If the two mirrors are accurately perpendicular to each other, the interference fringes will consist of concentric circles. Move M_1 to about 50 mm from the center of the beam splitter. Quite a large number of circular fringes should be seen on the screen and the fringes will be closely spaced. If M_1 is moved inward towards the beam splitter, gradually the fringe system will contract and fringes will disappear one at a time at the center of the pattern. As M_1 passes through the position of path equality (paths #1 and #2 equal) the central fringe will fill the entire field of view and no fringe will show on the screen. As M_1 is moved past the position of path equality (still moving towards the beam splitter) the fringes will begin to re-appear at the center and move outwards. Thus, the motion of the fringes is reversed, more and more fringes will become visible as the path difference is increased.

12. As mirror M_1 is moved a distance d, n fringes will either appear or disappear at the center of the pattern (or pass a given point near the center) when $n\lambda = 2d$, if λ is the wavelength of the laser light. If one counts fringes for a measured motion of the mirror M_1, the wavelength can be calculated. Since the wavelength of the Ne-Ne laser is precisely known, the mechanism for moving the mirror can be calibrated in terms of the known wavelength: $d = n\lambda/2$, where d = distance of travel of mirror M_1 in meters, n = number of fringes counted, $\lambda = 632.8 \times 10^{-9}$ meters for He-Ne lasers.

When the interferometer is used photoelectrically, the center of the fringe pattern is observed through a small pinhole and the fringe count is read off an electronic counter.

It must be noted that the interferometer described does not use an extended area diffused light source and the fringes observed here are not fringes of equal inclination. Nevertheless, the interferometer can be used for most applications in place of Michelson's original interferometer.

References:

T.S. Velichkina, O.A. Shustin and N.A. Yakovlev, Sov Phys Uspekhi
                                        v4 p523 (Nov/Dec '61)
Cenco Instrument Corp., Selective Experiments in Physics, "The Michelson
    Interferometer," No. 71990-548

U. Grigull and H. Rottenkolber, J Opt Soc Am v57 p149 (Feb '67)
E.M. Parma and B.C. Thompson, Am J Phys v39 p1091 (Sep '71)

""""""""""

Interference with Polarized Light
(Fresnel-Arago Laws)

A series of experiments performed by Fresnel and Arago in 1817 revealed the basic properties of polarized and unpolarized light. Only after these experiments had been performed was the transverse nature of light fully understood. The four interference laws of Fresnel and Arago can be easily demonstrated using a Michelson interferometer.

The interferometer is set up in the usual way. The two mirrors, M_1 and M_2 are adjusted to obtain straight line wedge fringes with a near zero path difference. Four linear sheet polarizers are needed. They can be introduced normal to the light rays at positions A, B, C, and D (see Fig. 160). A laser producing unpolarized light is used.

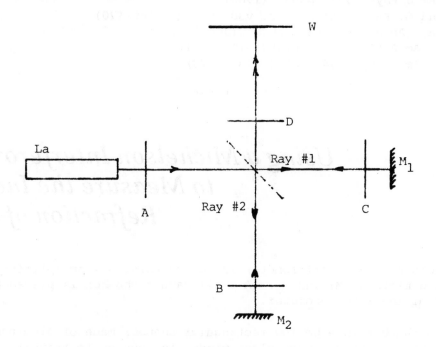

Fig. 160. Setup to demonstrate the Fresnel-Arago laws.

The interference laws of Fresnel-Arago are summarized below. Each law is demonstrated by appropriately placed polarizers.

Law No. 1:   Two rays, linearly polarized in the same plane, can interfere.

Demonstration: Polarizers placed at B and C. Their transmission axes parallel; fringes are seen on the screen (W).

Law No. 2:   Two rays, linearly polarized with perpendicular polarizations, cannot interfere.

Demonstration: Polarizers placed at B and C. Their polarization axes at right angles (crossed); fringes disappear on the screen.

Law No. 3:   Two rays, linearly polarized in perpendicular planes,
                   if derived from unpolarized light and subsequently
                   brought into the same plane, cannot interfere.
Demonstration: Polarizers placed at B, C, and D. B and C with their
                   polarization axes at right angles. D at 45 degrees to B;
                   fringes disappear.

Law No. 4:   Two rays, linearly polarized in perpendicular planes, if
                   derived from the same linearly polarized ray and subse-
                   quently brought into the same plane, can interfere.
Demonstration: Polarizers placed at A, B, C, and D. As in No. 3, but
                   A is added at 45 degrees to B (A and D parallel);
                   fringes can be seen.

References:

R. Hanau, Am J Phys v31 p303 (1962)
J.L. Hunt and G. Karl, Am J Phys v38 p1249 (Oct '70)
C. Pontiggia, Am J Phys v39 p679 (Jne '71)
E. Collett, Am J Phys v39 p1483 (Dec '71)
S. Mallick, Am J Phys v41 p583 (Apr '73)

"""""""""""

Using a Michelson Interferometer to Measure the Index of Refraction of a Gas

A Michelson-type interferometer is set up using a beam splitter and two
front surface mirrors, M_1 and M_2. A small vacuum chamber is placed in the
"fixed leg" of the interferometer..

The gas chamber is a hollow rectangular chamber made of aluminum. At
one end of the chamber a square glass window is mounted to transmit a laser
beam through the interior while retaining a good vacuum. At the other end is
mounted a front surface mirror (M_2) with a good vacuum seal. The other
mirror (M_1) is a gimballed mirror with precision adjustments.

As shown in Fig. 161, beams (1) and (2) are reflected by mirrors M_1 and
M_2 respectively. They are re-combined at the beam splitter (B) and the
fringe pattern may be projected on a screen. The two beams can be superposed
on the screen by making small adjustments to both mirrors until a fringe
pattern appears, as described in the previous exercise on a Michelson-type
interferometer.

Air in the chamber is removed with a vacuum pump. The valve is then
opened slightly to let air leak back into the previously evacuated chamber.
The air must be admitted slowly enough to permit counting of the interference
fringes as they move across the screen. There must be a reference point

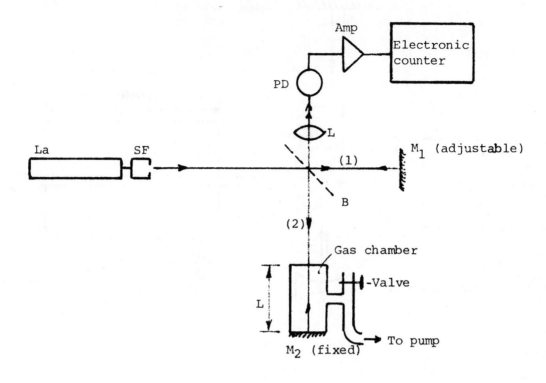

Fig. 161. Optical and electronic block diagram.

marked on the screen and each fringe is counted that moves past the reference. The refractive index n at a particular wavelength λ is computed from the relation: $n = (2L + N\lambda)/2L$, where

    n = index of refraction,
    N = number of fringes moving across the reference mark,
    λ = 6.328 x 10^{-5} cm for a He-Ne laser,
    L = inside length of the gas chamber in centimeters.

The optical path is 2L because light goes through the chamber twice. Light travels slower in air than it does in vacuum. This has the same effect as lengthening the optical path through the chamber. Thus, each fringe shift represents a virtual increase in optical path length equal to one wvelength of the laser light. By comparing the optical path length in the evacuated chamber with the apparently increased length when the chamber is filled with air (or other suitable gas) the index of refraction can be computed.

Counting of the fringes can be quite tedious and a strain on the eyes. A low-cost photo-detector/amplifier will provide a signal level output (0-5 V) which is sufficient to drive an electronic counter. A lens (L) of 36-50 mm focal length is used in the output beam of the interferometer to project the interference pattern on the detector. It is imperative that the dark fringes be at least the width of the active chip in the photodiode or the window diameter of the photo-detector.

All the components for the interferometer setup are included in the Metrologic Gas Index of Refraction Kit, No. 60-640. A base, beam splitter, precision gimballed mirror, gas chamber, vacuum pump, tubing, connectors and a sensitive air valve. A Metrologic No. 60-255 photo-detector/amplifier can be used to drive most of the Beckman-type electronic event counters.

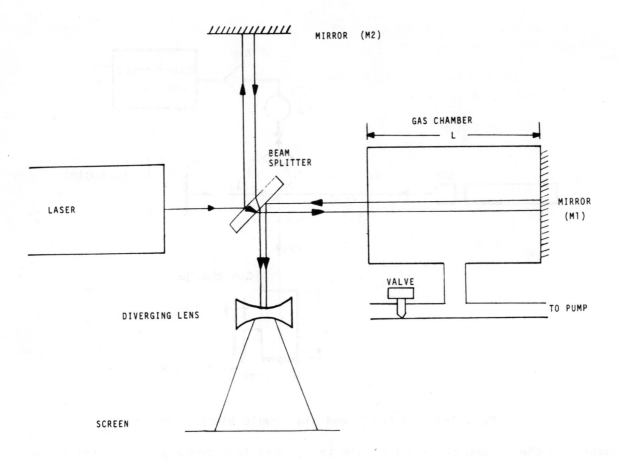

Fig. 162. Optical diagram of 60-640.

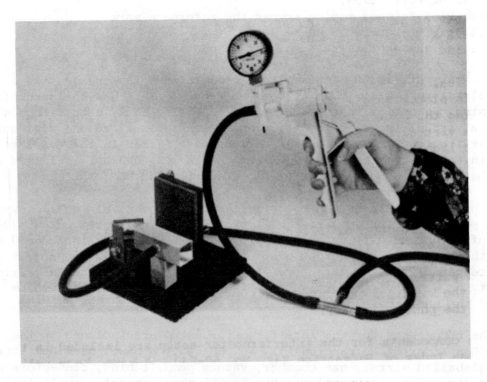

Fig. 163. Photograph of 60-640.

To measure the index of refraction of gases other than air, the following procedure is recommended:

1. With the equipment set up, connect the gas container to the rubber tube at the gas inlet valve.

2. Open the gas inlet valve of the 60-640 and close the valve on the supply of gas.

3. Pump the maximum vacuum (up to 25 inches of mercury vacuum in the gas chamber).

4. Open the gas supply valve, allowing the system to fill with the gas to be measured. The system is now primed, that is, air has been removed.

5. Close the 60-640 gas inlet valve and pump the maximum vacuum once again.

6. Observe the fringe pattern and slowly, very slowly, loosen the gas inlet valve screw until the fringes begin to move across the screen.

7. Count each fringe that moves past the reference mark on the screen. Continue to count until the fringes stop moving.

8. Use the formula to determine the index of refraction of the selected gas.

References:

F.Y. Yap, Am J Phys v39 p224 (Feb '71)
J.H. Blatt, P. Pollard and S. Sandilands, Am J Phys v42 p1029 (Nov '74)
K.O. Lindquist, Opt & Laser Technol p76 (Apr '72)

" " " " " " " " " "

Michelson Interferometer Used for Vibration Analysis

The vibrating surface under study is placed in an optical field consisting of two mutually coherent plane waves which propagate in somewhat different directions. If the surface is diffuse, a fringe pattern will be formed on it due to the interference of the two plane waves. If a time-averaged image of the surface vibrating within this interference field is formed (with a camera or by utilizing visual persistence of the eye), it is found that the visibility of the fringes formed on the surface is modulated by a function of the local amplitude of vibration. Thus loci of constant vibrational amplitude can be directly observed as regions of high or low fringe visibility in the image of the surface. Fringe visibility contrast can be enhanced by optical processing of the image.

The interfering plane waves are formed by a Michelson interferometer. When the spatial frequency of the fringes is low (i.e., in low amplitude sensitivity measurements), the amplitude mapping can be obtained by directly viewing the vibrating surface. When the fringe frequency is high, it is preferable to photograph the pattern and to perform simple optical processing on the transparency of the fringe pattern.

Figures 164 and 165 show the optical apparatus used for recording the fringe patterns and the system for optical processing.

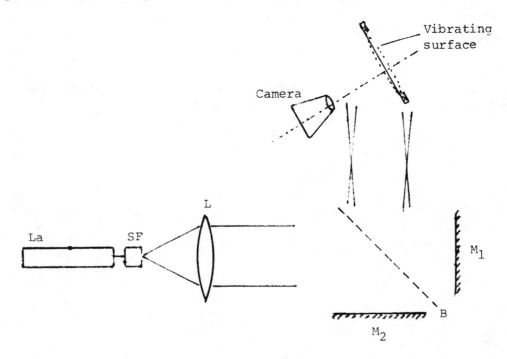

Fig. 164. Apparatus for analysis of vibrating surfaces.

Light from the laser is expanded, filtered and collimated. The collimated light is amplitude divided by a Michelson-type interferometer (beam splitter B, mirrors M_1 and M_2) into two plane waves traveling in slightly different

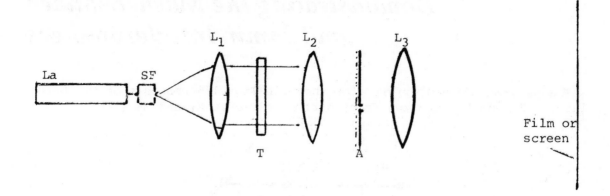

Fig. 165. Optical processor.

directions. The vibrating surface is placed in the optical field produced
by the interferometer and the resulting fringe pattern formed on the surface
is directly viewed or photographed. The vibrating object may be a circular
diaphragm excited by an air-coupled loudspeaker driven by an audio oscillator.
Low frequency fringes can be photographed on Kodak Plus X or Polaroid P/N film.
For high fringe frequency studies hologram recording material should be used.
(Hologram emulsions are described elsewhere in this book).

The optical processor (Fig. 165) is quite simple. A collimated beam of
laser light diffracted by a transparency of the fringe pattern (T) is Fourier-
transformed by lens, L_2. All spectral components other than one first-
diffracted order is filtered out by the properly placed aperture (A) and the
remaining wave is re-imaged by L_3 on to a photographic film. If the final
photographic print lacks contrast, it can often be enhanced by making a
xerographic print of it.

The technique described can be operated in a stroboscopic mode using a
shutter system or a modulated laser. This would allow one to observe the
amplitude structure at various intermediate phases of the vibration cycle.

Reference:

C.M. Vest and D.W. Sweeney, Appl Opt vll p449 (Feb '72)

" " " " " " " " " "

Demonstrating the Mach-Zehnder
and Jamin Interferometers

The Mach-Zehnder interferometer shown in Fig. 166 consists of two beam splitters, B_1 and B_2, and two front surface mirrors, M_1 and M_2.

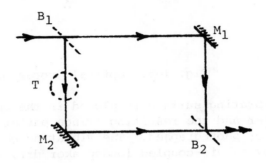

Fig. 166. The Mach-Zehnder interferometer.

The Mach-Zehnder interferometer is the only one suitable for studying slight changes of refractive index over a considerable area and is used extensively for measuring the flow patterns in wind tunnels. The light traverses a test region such as T in only one direction which simplifies the study of local changes of optical path in that region. A suitable adjustment of the mirrors causes the fringes to be localized in a chosen plane in the test are T. It is this ability to localize the fringes in a chosen plane which is the chief advantage of the Mach-Zehnder over the Michelson inter-ferometer. Alignment is made much easier if the mirrors are replaced by pentaprisms.

Figure 167 shows a simple demonstration setup.

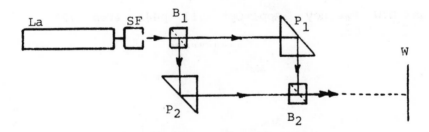

Fig. 167. Mach-Zehnder demonstration setup.

The four components are placed on a standard optical table. They can be kept in place with a small amount of putty, no component carriers are needed. Two 15 mm cube beam splitters (Edmund No. 30329) are used, and two 25x25x36 mm right angle prisms (Edmund No. 40995) take the place of the front surface mirrors. The screen, W, is placed about 1-2 meters from B_2. After the initial setup two bright light patches will be seen on the screen. These can be brought into coincidence quite easily. Using very light touch, P_1 and B_2 are fine-adjusted until fringes appear.

The Jamin interferometer is basically similar to the Mach-Zehnder design. It consists of two thick plane-parallel glass plates, GP_1 and GP_2. The back faces are reflecting (opaque aluminum or silver coated). The two plates are adjusted to be parallel to each other. Fig. 168 shows a plan of the instrument.

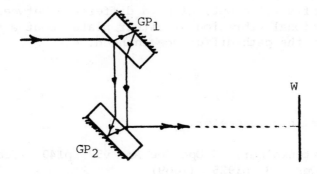

Fig. 168. The Jamin interferometer.

The incident ray is divided into two beams, one reflected from the glass surface and the other after refraction in the plate GP_1. These two beams arrive at another identical plate, GP_2, where they recombine to form interference fringes known as Brewster's fringes. If the plates are strictly parallel, the light paths will be identical. The fringes appear by slightly inclining one plate with respect to the other. If the glass plates are quite thick, the beams are sufficiently separated to be passed through two identical gas or liquid containers. If, as an experiment, we wish to measure the index of refraction of a gas at different temperatures and pressures, we start with two evacuated containers. Gas is slowly admitted to one of the containers. Counting the number of fringes crossing the field while the gas reaches the desired pressure and temperature, the index of refraction can be calculated as described elsewhere in this book.

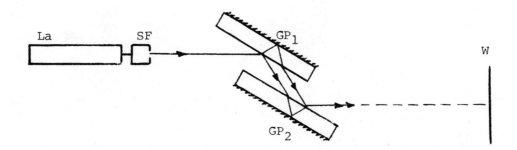

Fig. 169. Demonstrating Brewster's fringes.

Figure 169 shows the Jamin interferometer setup to obtain fringes of equal inclination. Two glass plates, GP_1 and GP_2, as nearly identical as possible and silvered on their rear surfaces, are required. They are placed on a rotatable optical table and held in place with a bit of putty.

If the faces of the plates are perfectly parallel the components of all rays interfere constructively to give a light patch of uniform intensity on the screen W, placed 1-2 meters from GP_2. If the plane of one plate is tilted through a small angle relative to the other, equally spaced fringes are formed. A lens, L, of short focal length may be placed in the path of the beams between GP_2 and W to obtain an enlarged interference pattern. The thickness of the plates may be between 5 and 25 mm. The important advantage of this Jamin-type interferometer arrangement is the steadiness of the interference pattern it produces. Translational motion of the two optical components does not affect the optical path difference between the interfering beams. Only by rotational vibration of either plate about a vertical axis through its center is the path difference changed.

References:

The Mach-Zehnder interferometer:

U. Grigull and H. Rottenkolber, J Opt Soc Am v57 p149 (Feb '67)
P. Hariharan, Appl Opt v8 p1925 (1969)
A.G. Kovaltchouk, R.G. Brzezenski and J.M. Bagarazzi, Am J Phys
v41 p1106 (Sep '73)

The Jamin interferometer:

M.V.R.K. Murty, Appl Opt v3 p535 (Apr '64)
G.A. Woolsey, Am J Phys v41 p255 (Feb '73)

" " " " " " " " " " "

Sagnac's Triangular Interferometer

The interferometer described here was introduced by G. Sagnac in 1910. In its original layout it was a "rectangular" interferometer. It was used by Michelson and Gale in their determination of the effect of the earth's rotation on the velocity of light. The "triangular" layout was used by Dowell to measure the length of end gauges. With spherical mirrors the triangular-path apparatus can be used for interferometric testing in wind tunnels. The grazing incidence triangular interferometer can be applied to check the quality of a granite surface for metrological inspections. An improved method for testing lenses, using an optical autocorrelator to measure the optical transfer function (OTF) of an imaging system, has been developed, - based on the Sagnac triangular interferometer.

Both the Michelson and Sagnac interferometers use beam splitters to divide the amplitude of the input beam to form the interfering beams. The Michelson instrument is better adapted to certain kinds of measurement where the two beam paths must be widely separated. The Sagnac instrument is easier to align than the Michelson interferometer and it works well with ordinary quality optical surfaces. Furthermore, the fringes from the Sagnac interferometer are more stable. Disturbances such as the vibration of a mirror, will have much less effect on fringe stability than with the Michelson interferometer.

Figure 170 shows the schematics of the Michelson interferometer (a); the rectangular Sagnac interferometer (b); and the triangular Sagnac interferometer (c).

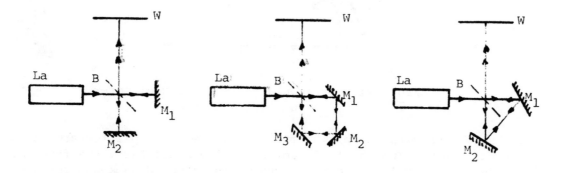

Fig. 170

It can be seen from Fig. 170 that by simply rotating mirrors M_1 and M_2 of the Michelson interferometer (a) we can form a triangular Sagnac interferometer (c). Both Sagnac configurations (b and c) are often referred to as cyclic interferometers, so called because the test and reference beams traverse their paths in opposite directions. In Figs. 170-b and 170-c the input laser beam is divided by the beam splitter (B) and travels clockwise, while the other is reflected and travels the equivalent path counterclockwise. If the instruments are perfectly aligned there is neither shear not path difference. The basis of the formation of fringes in the Sagnac interferometer is a slight misalignment of the mirrors, so as to produce varied paths for the two split beams going in opposite directions. Again, if the instruments are perfectly aligned and the beam splitter is turned slightly, then the counterclockwise path will be varied with respect to the clockwise path. The latter will be almost unaffected by the rotation of the beam splitter.

Figure 171 illustrates the experimental setup for the triangular Sagnac interferometer.

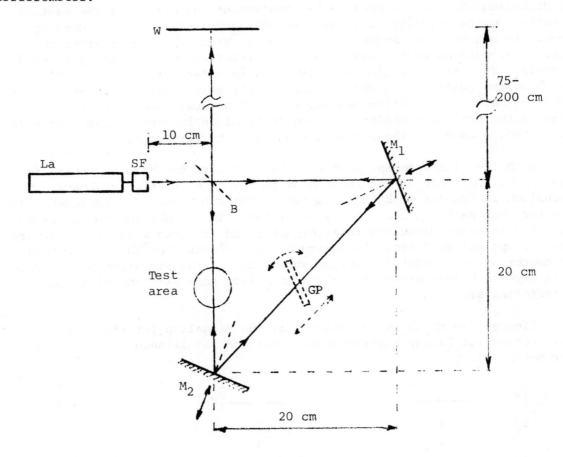

Fig. 171. Triangular Sagnac interferometer
(not drawn to scale).

The laser, lens-pinhole spatial filter and the beam splitter are set up on one optical bench. Two short (25 cm) optical benches carry the front-surface plane mirrors, M_1 and M_2. These two mirrors can be moved back-and-forth on the small individual optical benches as indicated by the arrows. The split beams travel at 90 degrees at the beam splitter. The incident and reflected beams enclose an angle of 45 degrees at each mirror. A small white card (filing card) is held in front of M_1. Two small, unequal sized light patches should be seen on the card. Manipulate M_2 to bring the patches close together to overlap. Next, put the white card in front of M_2 and adjust M_1 on the bench to bring the patches together as close as possible. This process must be repeated several times for optimum results. At this stage of the alignment, the component mount carrying M_2 is firmly affixed to its optical bench.

Fine adjustment of the beams is now carried out with M_1. This mirror is supported by a mirror tilter (for example, JODON MH-50, or Ealing No. 22-8825, or equivalent). Two micrometer screws which function as adjustments rotate the mirror about the horizontal and vertical axes in the plane of the mirror.

The clockwise and counterclockwise beams are re-combined at the beam splitter and they overlap on the viewing screen (W). The two beams are brought into coincidence with the help of the two micrometer screws on the mirror tilter.

If the micrometer screws do not have enough "reach" to bring the two light patches into perfect coincidence on the screen then the initial alignment of M_1 and M_2 was not carried out with proper care. Instead of repeating the whole procedure, come back with the micrometer screws to zero tilt on M_1 and rotate the beam splitter very carefully to a position at which the two light patches on the screen are very close together or overlap. Then work the micrometer screws again until a pattern of light and dark fringes appear on the screen. After the fringes are obtained, fine-adjust the laser carrier (up, down or sideways) to achieve an evenly illuminated area of fringes on the viewing screen. Initially it may take 15-20 minutes to align the instrument, but once the procedure is known it should not take more than five minutes to obtain fringes.

As mentioned previously, fringes are formed in the Sagnac interferometer by a slight misalignment of the mirrors. Once fringes, -any kind of fringes - are obtained on the screen, we leave the instrument alone and fine-tune the interference pattern with an "optical micrometer." The optical micrometer (GP), is a thin plane parallel glass plate (such as Kodak Slide Cover Glass). This glass plate is introduced in the beam path between mirrors M_1 and M_2, as indicated in Fig. 171 by dotted lines. The tipped glass plate produces further shear in the optical path and this may be added or subtracted from the shear introduced by the mirrors themselves. Move the glass plate along the optical path between the mirrors, tilt and rotate the glass plate slowly to change the fringe pattern. Then reverse the tilt and observe how the fringes change by reversing the direction of the shear introduced. Manipulate the glass plate to obtain fine and coarse fringes, vertical, horizontal and diagonal fringes.

Remove the glass plate from the beam path and replace it with a +1 or +2 diopter eye glass lens or similar long focal length (50-100 cm) simple lens. Move the lens along the optical path between the mirrors M_1 and M_2 and observe the formation of beautiful circular interference patterns.

Obtain good vertical fringes (with or without the fine-tuning glass plate), about 5 or 6 fringes in the field of view and place a glass cell (such as Ealing No. 22-8973) in the test area between mirror M_2 and the beam splitter B. Fill the cell with water and add a few drops of acetone (nail polish remover or film splicing cement). Observe the fringe pattern which readily shows the small changes in refractive index while gently mixing the liquids together. Instead of the glass cell, hold a hot soldering iron or a miniature candle in the test area and observe how the gridlike interference pattern is altered to a pattern that corresponds to the density variations in air produced by the heat of the soldering iron or the open flame of the candle.

References:

P. Hariharan and D. Sen, J Opt Soc Am v49 p232 (1959)
A. Zajac, H. Sadowski and S. Licht, Am J Phys v29 p669 (Oct '61)
Anon, Opt Spectra p22 (Jly '72)
C.L. Stong, Sci Am p112 (Feb '73)

The Point-diffraction ("Smartt") Interferometer

The Point Diffraction Interferometer (PDI) was first described in 1972 by Dr. Raymond N. Smartt of the University of Massachusetts. The PDI is a simple but powerful interferometric system.

When a beam of light is interrupted by any object in its path, part of the light is diffracted. If the object approximates to a point, the diffracted light forms a spherical wave. In the PDI, there is a minute precisely circular aperture (pinhole) in an evaporated thin film (gold or aluminum) on a rigorously non-scattering (mica) substrate. The thin film is semi-transparent. When the aperture is placed at the image formed by a lens, a mirror or an entire optical system such as a telescope, the diffracted light forms a reference beam, and interferes with the directly transmitted light. The resultant interference pattern is a direct measure of the quality of the optical system under test. The PDI differs from most other two-beam interferometers in that the light is divided into two components which produce a test and reference beam <u>after</u> passing through the system under test. The principle of operation of the PDI is illustrated in Fig. 172.

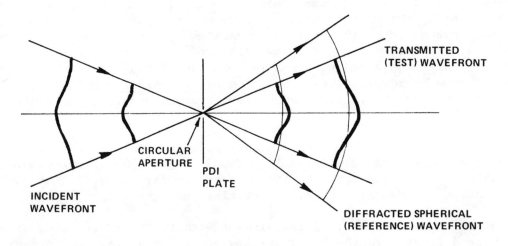

Fig. 172

The aperture is positioned at the image formed by the system under test. The direct wavefront is transmitted by the thin film with reduced emplitude. Some of the light is diffracted by the aperture to form the spherical reference wavefront. Interference between the two wavefronts occurs in the common region. The interference pattern provides a direct measure of the wavefront aberration introduced by the system under test. The fringes trace out contours of regions of equal optical path difference; successive fringes occur at increments of one wavelength of optical path difference. If the aperture is displaced along the axis away from the focal plane, circular fringes appear, since the wavefronts then have longitudinally displaced centers of curvature. Straight fringes are introduced when the aperture is displaced laterally from the axis.

The instrument shown in Fig. 173 consists of a delicate and carefully

Fig. 173. The Smartt PDI instrument (courtesy Ealing Corp.)

produced point diffraction plate mounted in a rugged microscope objective type
cell. This assembly is mounted in a unit equipped with three degrees of
translational freedom. On the rear of the unit is a detachable ground glass
screen. The screen provides a simple, ready mounted display surface for the
interferograms. The screen is easily removed for projection of the inter-
ferogram on to a larger screen. The interference fringes can be photographed
by removing the screen and using a 35-mm (SLR) camera focused to the aperture
of the lens under test.

A configuration of the interferometer to test a lens or equivalent optical system is shown in Fig. 174.

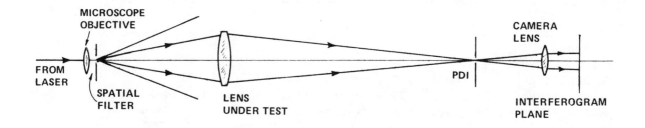

Fig. 174

The source is a low-power He-Ne gas laser. The beam is rendered divergent
by a high-quality microscope objective and spatially filtered. The PDI is
carefully centered on the point image formed by the lens under test. The
interference pattern is visible on the ground glass screen and the fringes can
be adjusted by means of the fine orthogonal adjustments of the interferometer
mount.

The PDI provides a convenient method to test camera lenses. By varying
the position of the lens in relation to the source, the lens performance can
be determined for objects at different distances, on and off-axis. The quality

153

of a camera filter can be determined simply by comparing the interferograms of the camera lens with and without the filter. The PDI is equally useful to test projection lenses and microscope objectives.

Another configuration of the interferometer to test a concave surface is shown in Fig. 175.

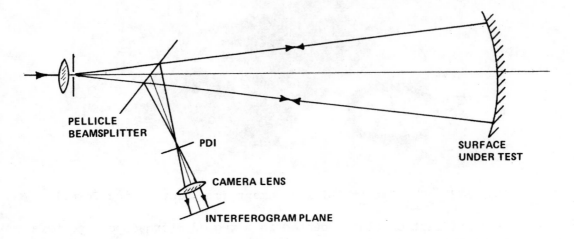

Fig. 175

A pellicle semi-reflector with a plane reflecting surface is used as the beam splitter. The PDI is adjusted as described above, to produce the fringe pattern, which can be observed directly, or photographed for quantitative analysis.

While the testing is more conveniently carried out when the surface under test has a reflective (aluminized) coating, an uncoated surface is adequate, fringe visibility remaining unchanged. Concave test plates as well as tele-scopic mirrors can thus be tested for sphericity during manufacture.

_____

References:

R.N. Smartt and J. Strong, J Opt Soc Am v62 p737 (1972)
R.N. Smartt and W.H. Steel, Japan J Appl Phys v14 suppl. 14-1 (1975)
"The Smartt Point Diffraction Interferometer," Instruction Manual,
    The Ealing Corporation, South Natick, Massachusetts, 01760.

" " " " " " " " " " " " "

The Fabry-Perot Interferometer

A Fabry-Perot interferometer consists of a pair of flat glass plates with semi-reflecting surfaces accurately parallel to each other and separated by distance d. An extended source of light, such as a frosted glass plate illuminated from the rear, is required to produce Fabry-Perot fringes. The concentric circular fringes are produced by the multiple reflection of light in the air film between the silver or aluminum coated surfaces.

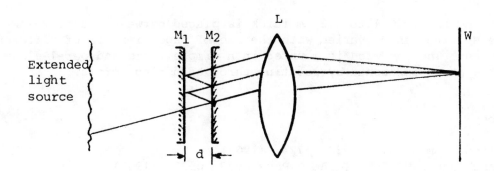

Fig. 176. Optical system of the Fabry-Perot interferometer.

Light is incident on M_1. Part of the light is transmitted directly through the system, but some of the light is reflected from M_2, then from M_1 again. In general, the outgoing beam is made up of light which has been reflected 0, 2, 4, 6... times and transmitted through two coatings, all summed together. Interference occurs when the transmitted rays are brought together by the objective lens. Destructive or constructive interference occurs depending on the path difference between adjacent rays. This difference is 2d cosϕ where d is the separation between the plates M_1 and M_2 and ϕ is the angle the particular ray makes with the normal to the plates. Bright fringes are visible in those directions where the path difference is an integral number of whole wavelengths. While there are only two interfering rays in the Michelson interferometer, there are many interfering rays due to multiple reflections in the Fabry-Perot interferometer. As a result, the Fabry-Perot fringes are much sharper than the Michelson fringes. When the distance d between the plates is increased, the circles expand, they move outward and new fringes appear in the center.

The simplest Fabry-Perot type device is a photographic cover glass or microscope slide, metallized on both sides by evaporation of aluminum in vacuum to produce semi-transparent mirror-like deposits. The ring system can be projected on a screen by focusing the laser beam with a short-focus lens (L = 30 to 50 mm f.l.) into a convergent beam and placing the metallized cover glass at the focal point. The viewing screen W is placed 3-4 meters away from the metallized glass plate. Since a good portion of the light is absorbed by the coatings, a laser of at least 2-3 mW output is needed for this demonstration (see Fig. 177).

Another simple system for demonstrating Fabry-Perot fringes consists of two pieces of good quality plate glass metallized on their facing surfaces.

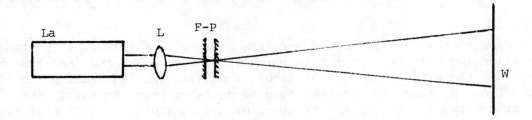

Fig. 177. Setup for projecting Fabry-Perot type fringes.

A paper ring spacer 0.1 to 0.2 mm thick is placed between the plates M_1 and M_2. The spacing can be varied with the use of paper inserts of different thicknesses. The sandwich is mounted in a simple frame and parallelism of the plates is accomplished by rotating three adjusting screws.

References:

W.A. Hilton, Am J Phys v30 p724 (1962)

A.P. French and J.H. Smith, Am J Phys v33 p532 (1965)

Central Scientific Co., "Selective Experiments in Physics," No. 71990-549
(1967)

Ya. E. Amstislavskii, Sov Phys - Uspekhi v11 p768 (1969)

"""""""""""

Demonstrations Illustrating
Temporal and Spatial Coherence

In many laser applications, such as interference and holography, the coherence properties of the laser are extremely important. The beam of light emitted by a laser can have the property of being almost completely coherent. In practice, light from a laser is not coherent. It is simply much more coherent than light from any other source. Just how much more coherent depends on the definition of the word "coherent."

The light produced by a laser can be thought of as a wave oscillating about 10^{14} times a second. For such a wave to be coherent two conditions must be fulfilled. First, the waves must be in step in the direction of propagation. This means that it must be of very nearly a single frequency (monochromatic), that is, the spread in frequency or bandwidth must be small. If this condition holds, the light is said to have high <u>temporal coherence</u>. Secondly, the waves must be in step in a plane perpendicular to the direction of propagation. This means that the wavefront must have a shape which remains constant in time. For example, a perfectly collimated beam of light has a plane wavefront. A point source of light emits a spherical wavefront. The light is said to be <u>spatially coherent</u> when the instantaneous phase difference between any two points in the radiation field is a constant. A perfectly coherent light source must be completely temporally and spatially coherent.

<u>Michelson's interferometer and temporal coherence.</u>

Figure 178 shows, in schematic form, a Michelson interferometer.

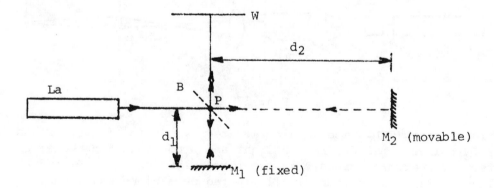

Fig. 178

The laser must operate in the TEM_{00} mode; in this mode the light is of constant phase across the beam and the pattern is perfectly circular. B is a beam splitter; M_1 and M_2 are front-surface mirrors. The two returning waves are recombined at P into a single beam which strikes the screen W or a photo-detector. The difference between the optical path lengths of the two waves interfering at the screen is $d = 2(d_2-d_1)$. Since M_1 and M_2 are at different distances (d_1 and d_2) from point P on the beam splitter, the resultant beam striking the screen will be composed of the original beam summed with a time-shifted version of itself. The optical path difference is increased by moving M_2 outward, away from P. When d becomes sufficiently large, the fringes seen will be of reduced contrast and with further increase of d, they will disappear. The disappearance of the interference fringes as this time-difference is increased gives the measure of the degree of temporal coherence.

The laser beam contains components with frequencies both above and below the light wave's nominal frequency: the beam has a finite bandwidth or spectral width.

Because the rate at which a wave changes phase is limited by its bandwidth, bandwidth (monochromaticity) is one measure of temporal coherence. The mean time between phase changes of the original beam is called the coherence time, τ_C. The coherence time is related to the bandwidth of the light wave, Δf, measured at the half-intensity points by $\tau_C = 1/\Delta f$. If the light wave travels with a velocity c, the distance over which the wave may be coherent is: $L_C = c\tau_C$ or $c/\Delta f$. This distance is called the coherence length.

Temporal coherence is also called time coherence and longitudinal coherence in the literature.

Young's double slit interferometer and spatial coherence.

Figure 179 shows a double slit interferometer.

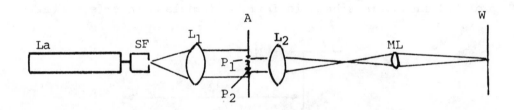

Fig. 179

The laser beam (TEM_{OO} mode) is collimated by spatial filter SF and lens L_1. Lens L_2 collects the diffracted light and produces a Fraunhofer diffraction pattern of aperture screen A in the focal plane of L_2. P_1 and P_2 are two small apertures in a screen A, illuminated by a plane wavefront. Apertures P_1 and P_2 act as sources in phase and interference fringes appear on the screen W. If P_1 and P_2 are moved progressively further apart in the collimated laser beam, the fringes on W get closer and closer together, become more and more diffuse and eventually disappear. In practice, several aperture screens are prepared for this demonstration, with spacing d ranging from 0.5 to 20 mm between apertures. To better observe the visibility of the fringes, a 20x to 40x microscope objective, ML, is used to project the fringes on the screen, W.

As mentioned before, when P_1 and P_2 are separated, the amount of interference decreases as the separation increases. The distance between P_1 and P_2 for which the degree of interference drops to some specified level is a measure of the spatial coherence of the light wave. A coherence area is defined as the area of the circle formed with points P_1 and P_2 at opposite ends of its diameter.

By moving the double slit aperture screen A across the expanded laser beam, it will be observed that as long as the collimated beam fills both apertures the fringe pattern remains visible and fixed, regardless of which portion of the beam the apertures occupy at any moment. This demonstrates that the laser light is spatially coherent across the entire beam. If a ground glass diffusing screen is placed between L_1 and A, the fringes would still be seen because the phase at the two apertures does not have to be the same, only of constant difference. If the ground glass were rapidly rotated then the fringes would not be seen because the light would become spatially incoherent.

Spatial coherence is also called transverse coherence in the literature.

Finally, interference by reflection from a thick plane-parallel glass plate is a good demonstration involving both temporal and spatial coherence. With an expanded laser beam very good straight fringes can be obtained on a nearby screen using a rather thick glass plate. The dependence of the fringe spacing and orientation on the reflection angle can be easily shown by tilting the plate relative to the laser beam.

References:

A.T. Forrester, Am J Phys v24 p192 (1956)

C.C. Cutler, Int'l Sci & Technol p54 (Sep '63)

G.B. Parrent, Jr. and B.J. Thompson, "Physical Optics Notebook," Society of Photo-optical Instrumentation Engineers, Redondo Beach, CA (1969) pp32+

M. Young and P.L. Drewes, Opt Commun v2 p253 (Nov '70)

S. Ezekiel, Opt Engineering v12 pG183 (Nov/Dec '73)

R.T. Pitlak, El-opt Syst Design v8 p23 (Sep '76)

"""""""""""

Laser Speckle Phenomena

<u>Random Interference Fields.</u>

Visible laser light scattered from a finely irregular surface or transmitted through a random inhomogeneous medium shows a speckled appearance caused by interference effects of the elementary waves emanating from each element of the scatterers.

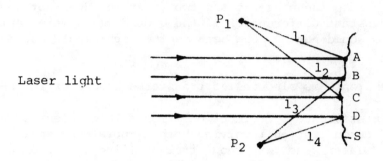

Laser light

Fig. 180.   Formation of speckles.

As shown in Fig. 180 visible laser light is incident on an optically rough surface, S, from which it is diffusely reflected. A, B, C, and D are scattering centers. At P_1 we observe a bright speckle if $l_2 - l_1 = \lambda n$ (n is an integer), corresponding to constructive interference. At P_2 we observe a dark speckle if $l_3 - l_4 = \lambda n + \lambda/2$, corresponding to destructive interference. Since a diffuse reflecting surface has completely random characteristics, the observed speckle pattern also has random characteristics.

Speckle has been a nuisance in coherent optical systems such as optical readers, optical processors and in the field of holography. A great amount of work has been done to reduce its disturbing effects. During the past few years, however, the speckle pattern has been recognized as a useful carrier of information rather than as a noise element. A variety of newer optical systems in which speckle is a useful entity include movement and vibration analysis, surface roughness measurement, non-destructive inspection devices, measurements of spatial and temporal coherence and analysis of eye defercts.

<u>Laser Speckle Effects.</u>

The basic arrangement for observing a speckle pattern is shown in Figure 181.

The laser beam is filtered with a lens-pinhole spatial filter. A non-specular surface (white matte paper or light colored wall) is illuminated with the "clean" divergent laser beam. If a Metrologic 60-618 spatial filter is used and the scattering surface is one meter away, a patch of light about 4 cm diameter will appear. The observer should be located near the laser and facing the scattering surface. The following effects will then be observed:

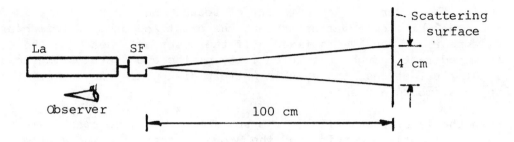

Fig. 181. Arrangement for viewing a speckle pattern.

1. The light scattered from the illuminated surface produces a speckled image. The illuminated area appears to be covered with bright and dark dots of a certain grain size.

2. The random interference field exists at all points in space between the scattering surface and the observer. That speckle is not localized at the surface is evident from the fact that it is always possible for the eye to see the granularity sharply, with or without spectacles, from any distance and any angle.

3. The apparent size or coarseness of the speckle is a function of the angular aperture or f/number of the optical system. As the aperture decreases, resolution of the optical system decreases, the granulation becomes larger. This variation in coarseness can be observed by interposing a variable iris or a set of pinholes of various sizes in front of the observer's eye and viewing the scattering screen with successively smaller apertures. The number of spots seen will be reduced and their apparent size or coarseness increased. The above observations were carried out about one meter from the screen..

4. When the observer moves back from the laser and the viewing distance increases to 2, 4, 6 meters from the screen the sparkling spots grow in size. Eventually single bright and dark patches cover the whole illuminated area. When the observer is careful to remain absolutely stationary, the pattern is stationary also.

5. When the observer slowly moves his head the speckle pattern appears to move too with respect to the screen. Close to the screen the spots move with the head. At greater distances from the screen some observers still see them move the same direction the head moves, while others see them move in the opposite direction.

6. The speckle pattern appears sharp to all observers, regardless whether they have emmetropic (normal), hyperopic (farsighted), or myopic (nearsighted) vision. The use of spectacles changes the focus of the optical system. This will change the distribution of the speckle pattern but it will not change its sharpness.

7. Most observers have difficulty keeping an object in focus when illuminated by laser light. The eye involuntarily tries to focus on the speckle. This is impossible since the speckle pattern is not located at any particular plane in space. When the eye focuses a little in front or a little

behind the object, it appears out of focus. When a newspaper is fastened over the white observation screen, the print and other detail blurs and disappears if the eye is fixed. If the observer's head (or the screen!) is moved fast enough, the retentivity of the eye averages the resulting light fluctuations, the speckle pattern disappears and the printing re-appears smooth and sharp.

8. When the laser beam falls on the surface of milk, no speckle is seen. Milk is a colloidal suspension and the Brownian motion of the scattering centers causes the interference field to move rapidly. Speckle is lost due to the persistence of vision.

9. The use of speckle pattern enables an unaided eye to recognize whether or not a surface is stable. Instability of a magnitude of 0.125λ can be resolved. While a human hand illuminated with laser light does not appear speckled, stable objects do have a frosty appearance.

References:

J.D. Rigden and E.I. Gordon, Proc IEEE v50 p2367 (Nov '62)
B.M. Oliver, Proc IEEE. v51 p220 (Jan '63)
C.C. Cutler, Int'l Sci & Technol p54 (Sep '63)
L.I. Goldfischer, J Opt Soc Am v55 p247 (1965)
J.N. Butters, Optics News (Opt Soc Am) p14 (Jan '76)

"""""""""""""

Dynamic Scattering Effect

This electro-optical effect was discovered in 1918 by Y. Bjornstahl. It was first discussed in depth by Heilmeier and his co-workers in 1968.

A liquid crystal cell operating in the dynamic scattering mode is essentially a parallel plate capacitor containing a liquid crystal dielectric. The plates are usually made of glass carrying transparent conductive coatings (NESA or NESATRON Glass by PPG Industries). The nematic liquid crystal (NLC) layer usually is 5-25 µm thick, has a resistivity of less than 10^{10} ohm cm and a capacitance of about 100 picofarads/cm^2. Thin films up to a few hundreds micrometers are quite transparent. Transparency can be further improved by orienting the NLC molecules perpendicular to the cell walls (homeotropic alignment). For a dynamic scattering device, liquid crystals with a negative dielectric anisotropy are used (such as MBBA). Licristal Nematic Phases #5A, #7A or #9A (products of E. Merck, Darmstadt) have wide operating temperature ranges (-10° to 75°C) and align themselves homeotropically with only a careful cleaning of the electrode surfaces. Dielectric conductivity is optimized for dynamic scattering. Another advanced nematic mixure is produced by Eastman Organic Chemicals, Cat. No. 14099, designated as "Dynamic Scattering III."

In an unactivated state the cell is quite transparent. When D.C. or a low-frequency A.C. field is applied to the electrode surfaces the orientation of the molecules is disturbed by the moving charge carriers. Above a defined threshold potential the current flow causes turbulence in the NLC layer which, in turn, causes a scattering of light due to the spatial variation in the index of refraction. The NLC layer loses its transparency and adopts the appearance of ground glass when observed in transmission through the transparent electrodes. Removal of the applied voltage allows the system to return to its original state as the surface orienting forces again take over. The cells work with voltages between 15-30 volts and dissipate approximately 100 microwatts of power per square centimeter of active cell area. Mechanically clamped (unsealed) cells when operated with continuous D.C. voltages have a lifetime of 10 to 100 hours. Sealed cells last several thousand hours when driven at 20-50 volts A.C. (60 Hz).

When a He-Ne laser beam illuminates a matte white screen or trans-illuminates a rear projection screen (ground glass), the illuminated area appears to be peppered with light and dark speckles. This "laser speckle" seriously degrades resolution of observed phenomena, such as interference and diffraction. Large cells in the dynamic scattering mode are excellent rear projection screens with a resolution of about 60 lines/mm. The scattering causes a wavefront impinging on a NLC dynamic scattering screen to be multiplied by a time varying random phase function. The laser speckle disappears because the scattering centers are constantly moving, thereby destroying any fixed phase relation in the scattered light (see Fig. 182).

The divergent laser beam emerging from the pinhole of the spatial filter (SF) is made to impinge on a biprism (BP). The interference fringes are observed on the activated NLC screen in transmission. (Do not look toward the laser unless the NLC cell is activated in the dynamic scattering mode!)

Laser light emerging from a dynamic scattering cell has an increased

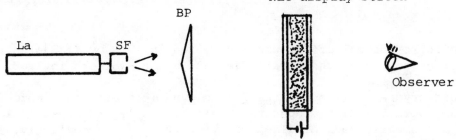

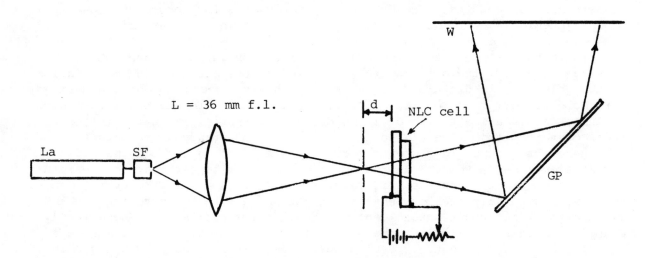

Fig. 182. NLC dynamic scattering screen.

line width (decreased temporal coherence) and a decreased spatial coherence.
The scattered field constitutes a source with very long coherence time. The
coherence time can be changed by changing the value of the applied voltage,
and the spatial coherence can be adjusted by varying the dimension of the
illuminated spot on the cell (Fig. 183).

Fig. 183. Setup for studying the coherence of
the scattered light.

The distance d is measured from the focal plane of lens L to the NLC film.
By changing d, the spot size on the NLC film is changed. GP denotes a thin
slide cover glass. The interference fringes are observed on a white cardboard
screen, W. The degradation of the degree of coherence of the laser beam can
be controlled with the applied voltage and by changing the distance d.

References:

G.H. Heilmeier, L.A. Zanoni and L.A. Barton, Appl Phys Lett v13 p46 (1 Jly '68)
G.H. Heilmeier, L.A. Zanoni and L.A. Barton, Proc IEEE v56 p1162 (Jly '68)
H. Mizuno and S. Tanaka, Opt Commun v3 p320 (Jly '71)
R. Bartolino, M. Bertolotti, F. Scudieri and D. Sette, Appl Opt v12 p2917
                                                        (Dec '73)
F. Scudieri, M. Bertolotti, R. Bartolino, Appl Opt v13 p181 (Jan '74)

Method and Procedure for Eye Testing with Laser Speckle

An expanded laser beam is projected on a diffusing screen and thereby a visual pattern (speckle) is produced. If relative movement between an observer and the diffusing screen is provided, the perceived movement of the speckle pattern is an indication of the observer's visual condition. The optical setup (Fig. 184) consists of a low-power Ne-Ne laser, a spatial filter to "clean" and expand the beam, and a means for diffusing the laser beam.

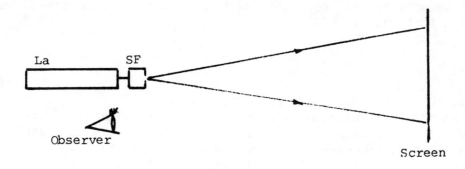

Fig. 184. Optical setup for eye-testing with laser speckle.

1. In a semi-dark room the expanded laser beam is projected onto a diffusely reflecting screen (white paper or light colored wall), placed 2-4 meters from the laser.

2. When the observer's head is moved extremely slowly from side-to-side while looking at the illuminated area on the screen, the speckle pattern appears to move.

3. If the movement of the speckle pattern is in the same direction as the relative movement between the observer and the screen (that is, if the observer's head moves to the right and the speckle pattern also appears to move to the right), the eye is hyperopic (farsighted).

4. If the movement of the speckle appears to be in the opposite sense, the eye is myopic (nearsighted).

5. If there is no speckle movement observed the eye is emmetropic (normal). A person who normally wears eyeglasses should take this test with and without the glasses.

6. In case of normal vision, myopia can be simulated by holding positive (convex) lenses of various powers in front of the normal or corrected eye.

7. To simulate hyperopia, negative (concave) lenses of various powers are held in front of the normal or corrected eye.

8. In case of astigmatism the apparent movement of the speckle pattern will depend on the refractive condition of the eye in the plane of the movement. For example, at some diagonal movement of the head the speckles will

remain stationary, up and down tilt of the head will produce speckle
movement in the opposite sense, and sideways motion will result in
speckle movement in the same sense. This example indicates mixed astigmatism, that is, the eye is myopic in the vertical meridian, hyperopic
in the horizontal meridian and emmetropic at some diagonal direction.

The above described setup can be used as a testing method for simultaneously screening entire groups. Since the speckle movement can be observed
at various angles and at various distances from the screen, many persons can
participate at the same time. While the subject looks at the laser illuminated area, positive and negative lenses of graduated powers are sequentially
placed in front of one of his eyes. The subject decides whether the speckles
move to the right, to the left or stand still ("null point"). Thus, this
method can be used to test children and illiterates who cannot read the
letters on a test chart.

References:

H.A. Knoll, U.S. Pat. No. 3,572,912 (30 Mch '71)
E. Ingelstam and S. Ragnarsson, Vision Res v12 p41 (1972)
W.N. Mohon and A.N. Rodemann, U.S. Pat. No. 3,724,933 (3 Apr '73)
W.N. Mohon and A.N. Rodemann, Appl Opt v12 p783 (1973)
S. Wittenberg, U.S. Pat. No. 3,792,918 (19 Feb '74)

"""""""""""

Vibration Detection Using Laser Speckle

A filtered and diverging laser beam illuminates the surface of a diffuse-ly reflecting, vibrating surface.

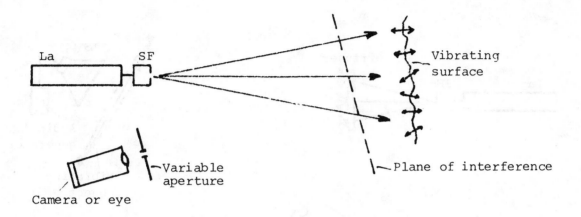

Fig. 185.   Interference displacement mapping system.

The light reflected from the surface consists of random phase waves which overlap in the region of space in front of the surface and form a speckle pattern.   If the eye or a camera is focused somewhat in front of the vibrating surface, the speckle pattern near areas that do not tilt appreciably during vibration will remain visible.   In areas of appreciable tilt the speckles become elongated and may be "washed out."   Antinodal areas can be detected by direct photography or a coherently illuminated vibrating surface.

References:

G.A. Massey, NASA Report No. NASA-CR-75643 (1965) and NASA-CR-985 (1968)
(NTIS, Springfield, VA 22151)

N. Fernelius and C. Tome, J Opt Soc Am v61 p566 (1971)

Examining Vibrating Objects by Means of Laser Speckle

With this simple method one can obtain data (visually or photographically) about the movement, or the state of rest, of various surface elements of a vibrating object to be examined.

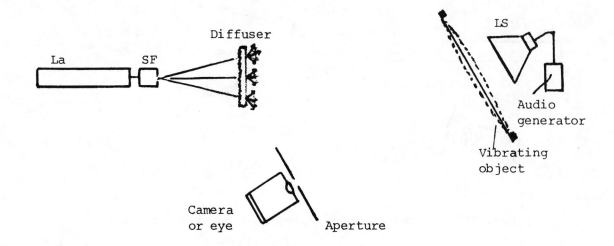

Fig. 186. Optical system for vibration analysis.

With reference to Fig. 186, the laser beam is expanded and filtered (SF) so that a luminous area is obtained on a ground glass diffuser. This area serves as a source of scattered light which illuminates the vibrating object to be examined. As an example, a circular diaphragm is set into vibration by the air-coupled loudspeaker which is driven by a variable frequency audio generator.

The vibrating diaphragm illuminated by scattered light from the diffuser can be viewed visually, or it can be photographed. By placing the diffuser between the light source and the vibrating object, a speckle pattern is projected onto the vibrating surface; i.e., the intensity pattern on the vibrating surface is due to the spatially random interference of the waves leaving the diffuser. Since the spatial variation of this speckle is random, it is blurred on surface areas that vibrate but yields a sharp, contrasty speckle pattern on nodal regions. The patterns observed on the vibrating diaphragm are reminiscent of the known Chladni figures. By photographing the pattern and developing the film with a suitable nonlinear process, the difference in contrast in the photograph can be enhanced.

The sensitivity of the method can be varied. The greater the ratio of the size of the luminous area to the diffuser-object distance, the finer becomes the speckle and the greater becomes the sensitivity. Thus, the sensitivity of the method can be varied within wide limits by altering the geometrical setup. The movement amplitude at which the transition from sharp speckle pattern to the homogeneous surface appears ranges from a few light wavelengths up to several tenths of a millimeter.

References:

B. Eliasson and F.M. Mottier, J Opt Soc Am v61 p559 (1971)
F.M. Mottier, U.S. Pat. No. 3,702,737 (Nov. 14, 1972)

""""""""""""

Speckle-shearing Interferometry

Speckle interferometry involves making a high-resolution double-exposure photograph of an object which is illuminated with coherent light. It is a simple method for measuring displacements. Since the approach is purely optical, there is no physical contact with the specimen. The fringes which depict the derivatives of displacements of the specimen are made visible by spatial filtering technique.

In the so-called speckle-shearing interferometry, two laterally sheared speckle images of the specimen are made to interfere. By double-exposure technique and with the specimen being deformed between the exposures, a fringe pattern is generated that depicts the gradient of the surface displacements of the specimen with respect to the direction of shear.

The optical system is shown schematically in Fig. 187.

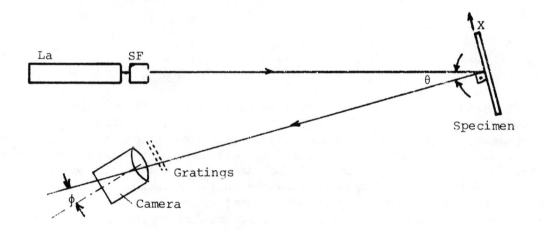

Fig. 187. Speckle-shearing interferometry.

The specimen is illuminated by an expanded laser beam making an angle θ with the y-z plane. The shearing interferometer consists of two identical diffraction gratings attached directly to the lens mount of the camera used to record the image of the test specimen. The grating frequency and the distance of the test specimen from the gratings are chosen so the angle of diffraction θ is large enough to avoid overlap of the first-order diffracted images with the directly transmitted image. The two gratings are in contact and with their rulings parallel to the y-axis; the two first-order images coincide and a single speckled image of the test object is formed in the film plane. If the two gratings are rotated in their own plane in opposite directions through equal small angles, two sheared images of the test object are obtained.

Speckle fringes are obtained by making two exposures on a photographic material with a highly nonlinear characteristic (such as Kodak Solar Recording Film SO-375), one before and the other after loading the test specimen. For a small shear, and if the angle θ at which the specimen is illuminated is close to zero, the speckle fringes give directly the gradient in the Y (shear) direction of the deflection in the Z direction. Bleaching the film is necessary where a low-power laser is to be used for data processing. A simple bleach would be 25 g $HgCl_2$ + 25 g KBr + 1 liter of H_2O; bleach 1 minute; wash 10 minutes; rinse in anhydrous methyl alcohol. After drying, the film is ready for the Fourier optical data processor. Note: With the double-exposure technique each exposure is one half of the total required.

In order to extract displacement data from the finished speckle transparency, the transparency is subjected to a simple optical processing system shown in Fig. 188.

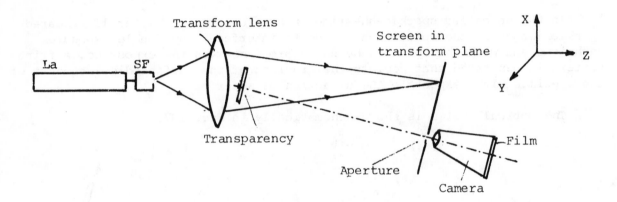

Fig. 188. Optical Fourier processor.

The laser beam passes through a lens-pinhole smoothing filter. The diverging beam is focused by the transform lens in the transform plane. The transparency of the speckle photograph is focused by the camera lens on the film plane of the camera. The camera aperture is located in the transform plane. The transparency holder and the camera are mechanically connected and rotatable about the center of the transparency. Rotation of this fixture will move the camera aperture away from the point where the laser beam is focused. By allowing two axes of rotation, the aperture can be translated in the X and Y directions.

The sensitivity of the displacement magnitude measurement depends upon the amount of offset of the camera aperture. The displacement component orientation depends upon the radial direction of the offset. The aperture of the camera lens transmits only a small portion of the frequency spectrum of the transparency into the camera. The camera lens performs an inverse Fourier transform on this segment of the spectrum. The image with its fringe pattern is recorded. The simplest and fastest method is to use Polaroid type 52 film.

The fringes obtained by spatial-frequency filtering are contours of constant displacement in the speckle photograph in the direction of displace-

ment of the camera aperture from the point of zero-order diffraction of the
laser beam.

_____

References:

K.A. Stetson, Opt & Laser Technol v2 p179 (1970)
C.L. Stong, Sci Am p106 Feb '72)
K.A. Stetson, Opt Engineering v14 p482 (1975)
G. Cloud, Appl Opt v14 p878 (1975)
P. Hariharan, Appl Opt v14 p2563 (1975)

"""""""""""

Image Plane and Focal Length Determination Using Laser Speckle

The phenomenon known as a speckle pattern results when laser light is reflected from or transmitted through an object having an optically rough surface.

When a narrow laser beam illuminates a ground glass the resulting speckle pattern can be observed on a screen located at a distance from the ground glass. If we observe the speckle pattern projected on the screen while the diameter of the illuminating beam is varied, we see a significant variation in speckle grain size on the screen. The effect can be utilized to determine an image plane location or the focal point of an optical system where direct observation of the beam waist on a screen is difficult because of the small size.

Figure 189 shows the schematic of the optical system to determine the location of the image plane (I) of a lens system (L).

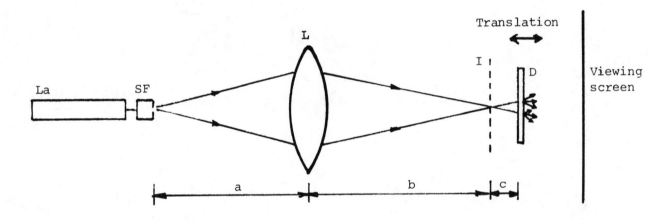

Fig. 189. Optical system to detect image plane.

The laser beam is filtered and made divergent with a lens-pinhole spatial filter (SF). The lens system (L) images the pinhole onto the diffuser (D) such as a fine satin-finished ground glass, or, preferably, a sheet of ordinary antiglare glass. Light scattered by the diffuser produces a speckle pattern on a viewing screen positioned at a convenient, arbitrary distance (c) from the diffuser. As the longitudinal position of the diffuser is altered and the image plane approached, there is an enormous increase in the size of the speckle on the screen while the decrease in the beam width on the diffuser is barely observable. Thus, it is much easier and more accurate to adjust the diffuser for maximum speckle size to find the image plane than to observe the beam waist directly.

If the focal length of a lens has to be determined, the same setup can be used. Once the image plane is found, the object distance (a) and image distance (b) can be measured. The focal length can be obtained from the formula:

$$1/a + 1/b = 1/f \qquad \ldots\ldots(1)$$

We transpose Eq. 1 by solving for f,

$$f = (a \times b)/(a + b) \qquad \ldots\ldots(2)$$

The author developed an opto-electronic focusing system to replace the subjective visual observation of speckle patterns. The system is illustrated in Fig. 190.

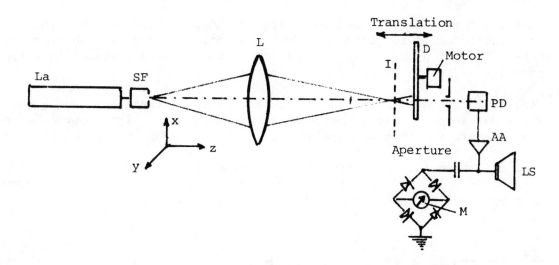

Fig. 190. Opto-electronic focusing system.

The viewing screen of Fig. 189 is replaced by a photo-detector (PD). The signal output is fed to an audio amplifier (AA) and then to a loudspeaker (LS). It will be found by longitudinally moving the diffuser (D) along the Z-axis that a weak, high-pitched whistle is emitted by the loudspeaker. The whistle becomes louder when the diffuser reaches the image plane, then the volume drops again as D is moved past the best focus point. Next, the diffuser is translated across the light beam in the X-Y plane. Quick back-and-forth movement of the diffuser causes the speckle pattern as a whole to move across the aperture of the photo-detector, giving an output signal proportional to the intensity distribution of the pattern. Instead of using a lateral vibratory motion, the diffuser (a ground glass disk) is mounted on the shaft of a small D.C. motor (for ex. Edmund No. 41,064) operating between 4 and 6 volts. The rotating disk is then introduced into the light path and slowly moved back-and-forth along the Z-axis. The maximum audio output is obtained when the rotating diffuser is in the image plane. As the plane of the diffusing surface departs from the image plane the audio signal first rapidly then gradually decreases in volume. The amplified audio signal may be coupled through a capacitor to a full-wave rectifier. The rectified audio signal is applied across meter (M) to give a comparative sound-level reading.

References:
W. Martienssen and E. Spiller, Naturwiss v52 p53 (1965)
T. Sawatari and A.C. Elek, Appl Opt v12 p881 (1973)
J. Ohtsubo and T. Asakura, Nouv Rev d'Opt v6 p189 (1975)
A.F. Fercher and H. Sprongl, Opt Acta v22 p799 (1975)

The Optical Doppler Radar

In 1842 Christian Johann Doppler, the Austrian mathematician and physicist advanced the theory: The frequency of a radiation will shift if its source or receiver is in motion. Doppler's theory was first verified in acoustics with the discovery that the speed of motion of a sound's source or its receiver, bore a direct relationship to the speed of the sound's propagation. That familiar change in a sound's pitch when its source or hearer is moving is known now as the Doppler effect.

The Doppler effect is used in microwave radar systems for the determination of the velocity of moving targets. Here the reflected signal is beaten against a reference signal produced by a local oscillator. The difference in frequency is proportional to the velocity of the target.

When the method of detecting small frequency differences of two light waves was developed (optical photo-mixing), and coherent light sources were invented (lasers), it became possible to construct an optical Doppler radar.

Let us assume that a collimated laser beam of frequency f and wavelength λ is incident normally on a front-surface silvered plane mirror (the reflector) which is approaching the source with velocity v (see Fig. 191).

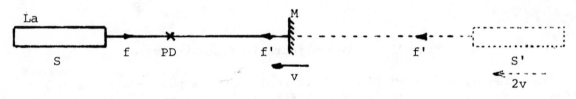

Fig. 191

The reflected beam comes from a virtual source S' moving towards S at velocity 2v, with a frequency f'. A photo-detector (PD) receiving both primary and reflected waves detects the beat frequency $f' - f = \Delta f = f \times 2v/c$. Since $f = c/\lambda$, therefore $\Delta f = 2v/\lambda$.

Consider now the path difference between S and S' and the fixed point PD, the photo-detector. The primary and reflected waves interfere. The path difference changes by λ for a movement of $\lambda/2$ of the mirror. Thus, the rate of movement of fringes past the reference point (PD) is equal to twice the velocity of the mirror measured in units of wavelengths per second.

To demonstrate the Doppler shift from a moving mirror, a Michelson interferometer assembly (Fig. 192) is used.

The use of corner cube reflectors rather than plane mirrors greatly simplifies the demonstration since the former need only an approximate lateral alignment. The detector is a photo-meter, a photo-tube, or a photo-diode of sufficient sensitivity and frequency response. (For example, Metrologic 60-255 or 60-230).

The output of the photo-detector is connected to the input of an amplifier. The output of the amplifier is fed to a loudspeaker (and an oscilloscope, if

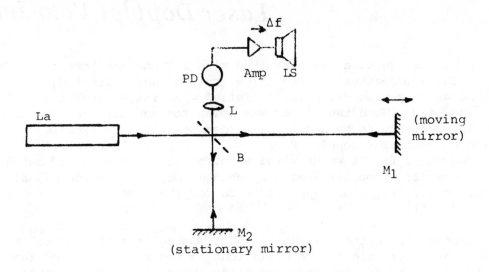

Fig. 192

desired). One of the mirrors, M_1, is mounted on a screw-driven optical bench
attachment which is driven manually or with a small motor to produce a high-
pitched audio note. The Doppler 'whistle' produced by slowly moving the mirror
is plainly audible. If an oscilloscope is connected to the amplifier output,
it provides a simple means of measuring Δf (the Doppler frequency) and since
the wavelength of the laser light is known, to calculate the speed of the
moving mirror.

Rapid movement of the mirror will produce frequency changes greater than
20 kHz which are supersonic and cannot be heard. If we wish to calculate the
velocity of the mirror for a Doppler change in frequency $\Delta f = 1,000$ Hz, the
required value of v is: $v = \Delta f \lambda / 2 = 1,000 \times 633 nms^{-1}/2 = 3.16 \times 10^5 nms^{-1} =$
0.316 nm s^{-1}.

This velocity of about 1/3 mm per second is easily obtained with the micrometer screw moving mirror M_1.

If the apparatus used in the demonstration is set up on ordinary tables,
there is usually enough random vibration of the mirrors to produce a low-
frequency 'rumble.' If the apparatus is used in artificial light, excessive
hum may result unless the photo-detector is shielded from room illumination.
When mirrors M_1 and M_2 are set close to zero path difference, the amplifier
gain is sufficient to produce acoustic feedback to the mirrors, causing
'ringing' and continuous oscillation at zero path difference. This effect
can be used to find the position of M_1 for zero path difference!

Reference:

P.W. Fish, Phys Educ p20 (Jan '71)

" " " " " " " " " "

Laser Doppler Velocimetry

The previous exercise was performed with a Michelson interferometer assembly. The loudspeaker emitted a low 'rumble' when mirror M_1 was moved slowly, and an ever higher Doppler 'whistle' when faster and faster mirror movement was used. When the mirror was moved too rapidly, the signal became too high in frequency to be heard. This demonstrated the excellent velocity resolution of the Laser Doppler Radar, which is directly proportional to the carrier frequency, f. If we know the velocity of the mirror and match the beat frequency (Δf = Doppler frequency change) on a calibrated signal generator, we can obtain a fairly close approximation for the wavelength of the laser light ($\lambda = 2v/\Delta f$).

In essence, the method of optical photomixing is analogous to the super-heterodyne method of detecting oscillations in the radiofrequency range. This method permits us to detect frequency differences of two light waves down to 1 Hz. In practice, this means we can detect the motion of an object having a velocity of 1μm/second; - for example, plant growth, or the motion of glaciers, or particles in Brownian movement, etc.

A very interesting demonstration can be set up to show the sensitivity of the laser-Doppler system. Again the Michelson interferometer assembly is used. After alignment, the movable mirror, M_1 is removed from the optical bench. In its place a piece of "Scotchlite" brand retroreflective tape is held so as to intercept the beam. The slightest movement of the tape along the beam path will produce a whistle in the loudspeaker very much like the whistle heard when a superheterodyne radio is fine-tuned. The effect is so sensitive that Doppler 'squeal' can be produced merely by holding one's hand, or a piece of white paper, in the beam a few centimeters away from the beam splitter. The reason for this effect is that a small percentage of the back-scattered light returns along the laser beam axis to the beam splitter where it coincides and interferes with the beam coming from the fixed mirror, M_2. The back-scattered light is changed in frequency by the amount of the Doppler shift caused by the slight trembling of the hand or hand-held piece of paper.

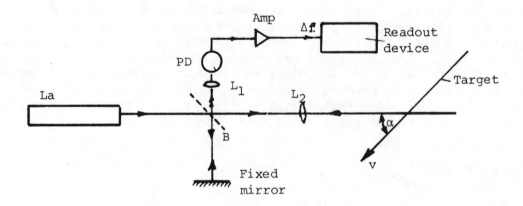

Fig. 193. A laser-Doppler system.

One of the most important advantages of the laser-Doppler technique is that the moving target is not affected since the 'probe' is a beam of light.

The laser-Doppler principle can be used to measure the speed of a wide variety of moving targets. Figure 193 shows the system which will work for strong back-scattered signals.

The setup is a Michelson interferometer configuration. The moving mirror M_1 is replaced with a moving target and the amplified Doppler shift is read off on a spectrum analyzer or measured by means of a frequency discriminator, or a frequency tracker. The velocity of the target may be determined from the relationship: $\Delta f = (2v \cos \alpha)/\lambda$, where Δf is the Doppler change in frequency, λ is the wavelength of the laser light, and α the angle between the scattered beam and target direction.

In many applications the scattered signal is very small because of the nature of the target and/or the angle between the laser beam and the target movement. Since the target is a given factor, one has to keep α as large as possible to maximize the light back-scattered towards the beam splitter. On the other hand α must be less than 90 degrees, at which angle $\cos \alpha = 0$. The choice of α is therefore a compromise between these factors. Lens L_1 is used to concentrate the interfering light beams on the photo-detector (PD), and lens L_2 focuses the laser beam on the target and it also collects the back-scattered light.

The laser-Doppler method of measuring velocities can be succesfully applied in various fields of science and industry. In a typical gas flow application, the flow is seeded with micrometer-size solid particles (such as smoke particles or latex spheres) to enhance the intensity of the scattered light. The Doppler shift is determined by causing the scattered light to heterodyne, or beat, with unshifted radiation from the same source. This scheme, -as already described, - is difficult to align and efficient heterodyning (that is, the production of a strong signal at the difference frequency) is obtained only if the wavefronts of the two beams have nearly the same curvature. Typically, the alignment must be better than one arc minute when visible laser light is used.

At this writing, there are about two dozen schemes devised and described in the literature which are much simpler to align. In general, the optical setups for laser-Doppler-velocimeters (LDVs) are quite simple. Some are more suitable for forward-scattered light, while others function with back-scattered light. All these various designs can be classified into three general categories: 1) local oscillator heterodyne,   2) differential heterodyne, and 3) symmetric heterodyne.

Figure 194 shows the schematic of a local heterodyne arrangement.

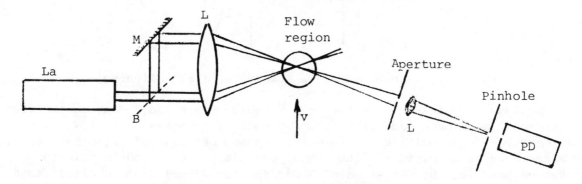

Fig. 194. Goldstein-Kreid system.

Figures 195-a and 195-b show schematics of differential heterodyne arrangements.

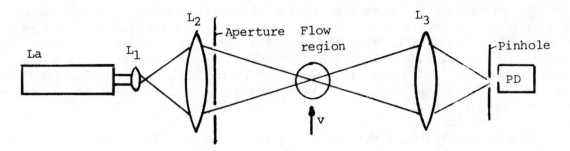

Fig. 195-a.   The Rudd system (self-aligning).

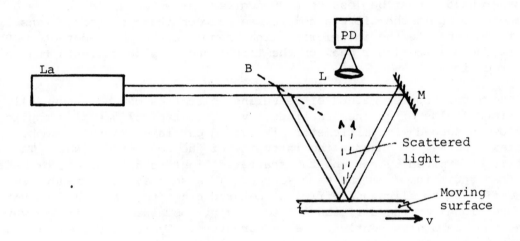

Fig. 195-b.   The Penney system.

Figure 196 depicts the schematic of a symmetric heterodyne arrangement.

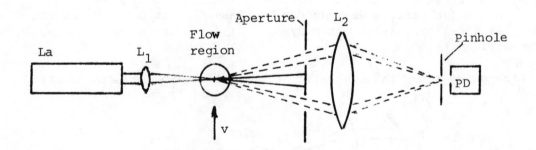

Fig. 196.   The Mazumder-Wankum system (self-aligning).

The laser-Doppler-velocimeter is now twelve years old.   Many scientific measurements have been made with LDVs for control purposes.   The LDV may not only measure the velocities of fluids and gases, but also of solid materials. LDVs can be used in strip rolling, paper and glass making and in printing. It can be used for the detection of clear air turbulence from an aircraft in flight, or the measurement of blood flow in human arteries.   A short bibliography follows for those readers who wish to undertake an in-depth study of this fast growing field. Some of the more recent papers (published after 1970) include references to more than 160 related articles.

Suggested Additional Readings:

Y. Yeh and H.Z. Cummins, Appl Phys Lett v4 p176 (1964)

J.W. Foreman, Jr., E.W. George, R.D. Lewis, Appl Phys Lett v7 p77 (1965)

R.I. Goldstein and K. Kreid, Trans ASME ser. E. v34 p813 (1968)

M.K. Mazumder and D.L. Wankum, IEEE J Quant Electron vQE-5 p316 (1969)

C.M. Penney, IEEE J Quant Electron vQE-5 p318 (1969)

S.J. Ippolito, S. Rosenberg, M.C. Teich, Rev Sci Instrum v41 p331 (1970)

P.S. Bedi, J Phys E: Sci Instrum v4 p27 (1971)

M.J. Rudd, Opt & Laser Technol p200 (Nov '71)

K.A. Blake, J Phys E: Sci Instrum v5 p623 (1972)

R. Manning, Opt & Laser Technol p114 (Jne '73)

C.P. Wang and D. Snyder, Appl Opt v13 p98 (1974)

B.S. Rinkevicius, Sov Phys - Uspekhi v16 p712 (Mch-Apr '74)

F. Durst and D.H. Stevenson, Appl Opt v15 p137 (1976)

R.V. Mustacich and B.R. Ware, Rev Sci Instrum v47 p108 (Jan '76)

D.K. Kreid and D.S. Rowe, Appl Opt v15 p321 (1976)

D.K. Hutchins, Am J Phys v44 p391 (Apr '76)

D.F.G. Durao and J.H. Whitelaw, Contemp Phys v17 p249 (May '76)

L.H.J. Goossens and J.A. van Pagee, J Phys E.: Sci Instrum v9 p554 (Jly '76)

W.M. Farmer, Appl Opt v15 p1984 (Aug '76)

M.J. Lalor, and W. Weston, Phys Educ v11 p425 (Sep '76)

""""""""""""

Point-of-Sale Label Reading

The technical term 'point-of-sale' (POS) means the time and place at which the customers hand over their money to the cashier in a retail establishment, or the checkout counter in a supermarket.

More and more supermarkets are equipped with electro-optical checkout stations that read, enter and register the necessary information as the cashier bags the merchandise.

A full supermarket point-of-sale system consists of a laser scanner head and a detector head at each checkout station. These electro-optical terminals act as electronic cash registers. The output of each terminal is fed to a central computer which carries in its memory all data vital to pricing, taxes and inventory.

The scanning of information from the labels of supermarket merchandise is necessary for a supermarket system. This information is encoded in a standard form, the Universal Product Code (UPC) symbol that is already being printed on more than 80% of the supermarket products (see Fig. 197). The basic characteristics of the symbol are the rectangular array of black and white bars which can be read by a scanner and at the bottom of each symbol two sets of numerals representing the human readable characters. The left-hand set of numerals forms a number that identifies a product manufacturer. The right-hand set of numerals forms a number that the manufacturer uses to identify a particular product.

Fig. 197. Oversize UPC symbol.

Each character of a code is independent and is represented by two black bars and two white bars (spaces) of varying widths for reading by a symbol reader, which is embodied into the checkout counter. As the product is swept over a small window, flush with the counter, the package is briefly illuminated by a sharply focused scanning beam of light from a low-power He-Ne laser. The scanning laser beam head interrogates the symbol printed on the package. This is done by monitoring the diffusely reflected light contrast variations from the black and white bars. The detected signal variations are then processed and passed on to the computer which in turn relays pricing and tax information to the cash register at the checkout station.

As the scanning beam moves from light to dark areas, the photo-detector output voltage goes from high to low. A signal processor and digitizer separate the signal from background and other interfering light and digitize the signal into a series of 1's and 0's, corresponding to light and dark areas of the printed UPC symbol. The binary information is fed to the computer to identify the product, and to interrogate the computer memory for the current product price. A completely automatic POS system displays and produces in human-readable form a checkout slip at the cash register. The slip gives not only the price but also the name of the purchased product, the tax collected and the grand total. One of the most valuable function of the automated POS system is that it provides current inventory data to a central computer of a supermarket chain.

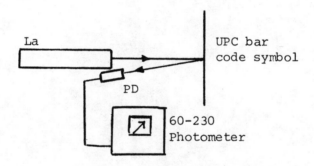

Fig. 198. Simple code reader.

Masking tape or putty is used to attach the photo-detector (PD) of Metrologic's 60-230 Photometer to the front end of the laser. The angle of the photo-detector is so adjusted that it will detect a beam reflected from a distance of 10 cm (4") in front of the laser. The laser beam is swept across the oversize UPC symbol (Fig. 197) and the electrical signal is read on the meter.

Figure 199 shows a simple mechanical scanning system. A convex lens (not shown) is used to project the laser beam to a tiny oscillating mirror and on from there to focus it on the UPC symbol. The vibrating mirror provides the scanning motion. Figure 200 shows an electrical toothbrush with the mirror cemented to the back of the brush attachment. A very thin front-surface mirror is used. The photo-detector (PD) reads the reflected signal. The amplified output from the photometer (60-230) is fed to an oscilloscope so that the 1's and 0's can be read.

(see illustrations on next page)

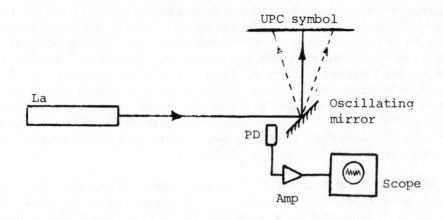

Fig. 199. A simple mechanical scanning system.

Fig. 200. Vibrating toothbrush with mirror.

References:

"UPC Symbol Specification, January, 1975," published by Distribution Codes, Inc., Alexandria, Virginia.
"Symposium on the Universal Product Coding System," U.S. Senate Hearing, Dec. 11, 1974 (Serial No. 93-114), Superintendent of Documents, U.S. Government Printing Office, Washington, D.C. 20402 (1975)
D. Savir and G.J. Laurer, IBM Syst J n1 p15 (1975)
G.F. Marshall, Optics News (Opt Soc Am) p5 (Jan '76)
G.F. Marshall, Phys in Technol v7 n4 p141 (Jly '76)
B.A. Young, Opt Engineering v15 p371 (Jly-Aug '76)

"""""""""""

Interference in Scattered Light

Most interference experiments demand careful adjustment of the apparatus. There is a very simple way to obtain beautiful interference fringes with the help of a He-Ne laser. The following experiment utilizes two luminous point sources derived from the same beam by means of scattering and diffraction. The setup is shown in Fig. 201.

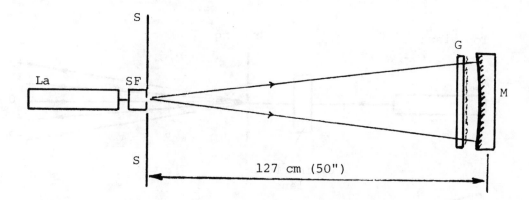

Fig. 201. Interference in scattered light.

The pinhole of the spatial filter (SF) is at the center of curvature of a concave telescope mirror (M) silvered on its front face (Edmund No. 40,913, 6 cm diameter, 25" focal length). A plane parallel plate of glass (G=Kodak Slide Cover Glass) is lightly oiled on one face and is then rubbed with a piece of cloth. Lycopodium, baby powder, chalk dust or rice flour is then dusted over it so that the particles of powder are deposited fairly uniformly on the glass plate. A fine-textured scattering film is thus obtained. The nature of the particles does not affect the geometry of the observed interference pattern, but may influence its brightness. The dusted glass plate is placed in front of the concave mirror and the rings are seen extending upon the viewing screen S-S. The dusted side of the glass plate should face the silvered front of the mirror, keeping the two apart by thin strips of paper inserted at the edges. The closer the plate is to the mirror the larger the size of the concentric rings on the screen. Each scattering particle and its reflection together act as a double source of scattered light. Interference from each twin source produces the basic fringe pattern. This is the same for all scattering particles and, thus, the composite interference pattern formed by the mirror is the superposition of all the geometrically identical patterns.

References:

A.J. De Witte, Am J Phys v35 p301 (1967)
E. Hecht, Am J Phys v43 p714 (1973)
C. Pontiggia, and L. Zefir, Am J Phys v42 p692 (1974)

"""""""""""""""

Demonstrating the Paths of Diffracted Rays

This setup shows ray paths in a plane, not just spots on a screen, representing the different orders of diffraction from a grating. The diffraction angles of the various orders can be measured directly.

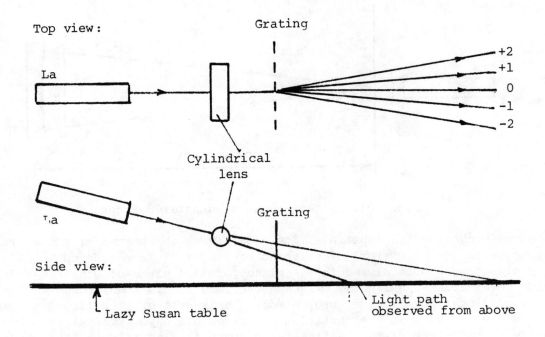

Fig. 202. Demonstrating diffracted rays.

" " " " " " " " " " "

The Poisson-Arago Spot

From Fresnel's wave theory of light Poisson deduced in 1818 that a spot of light should appear at the center of the shadow of a circular obstruction. Poisson's predicted spot had been observed by Maraldi one hundred years earlier. Arago performed Poissson's thought experiment and rediscovered Maraldi's long forgotten spot. This phenomenon is one historically imprtant confirmation of Fresnel's wave theory of light and it is known as the Poisson-Arago bright spot.

The Poisson-Arago spot can be made visible in the laboratory with the aid of a He-Ne laser. Ball-bearings (BB) are used as the round objects on which the diffraction of the laser radiation takes place. The diameter of the ball is about 2 mm. The ball is glued on to the surface of a +3 diopter (L = 333 mm f.l.) lens. The lens, and the ball fixed to it, is placed in the divergent laser beam approximately 30 cm from the spatial filter (Metrologic 60-618).

The magnified diffracyion pattern is observed on a screen (W) situated at a distance of 4-6 meters from the lens.

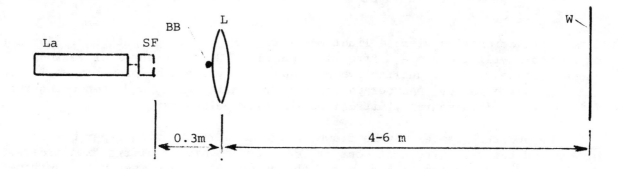

Fig. 203.   Setup to demonstrate the Poisson-Arago spot.

The Poisson-Arago spot can be clearly observed at the center of the geometrical shadow of the ball on the screen. Beyond the region of the geometrical shadow the outer diffraction rings are visible.

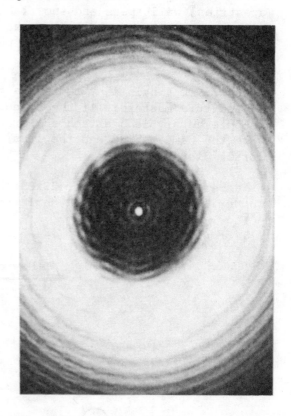

Fig. 204.   The Poisson-Arago   Fig. 205.   Enlarged center portion
              spot.   of Fig. 204.

References:

M.E. Hufford, Phys Rev (2) v3 p241 (Apr. 1914)
H. Schober, and R. Krusche, Optik v30 p314 (1969)

Making Photographic
Diffraction Apertures and Masks

Diffraction with visible light produces some beautiful patterns. Any sort of aperture will produce a diffraction pattern. Good apertures of various shapes can be made for qualitative and quantitative work by preparing a black-on-white master, photographically reducing its size, and then using the film negative (transparency) itself as the diffraction target.

The aperture master may be drawn in ink on white drawing paper but ink bleeds and the resulting soft contours destroy the higher diffraction orders. Much better results can be obtained with the help of the press-type dry-transfer sheets obtainable in art supply stores. A large variety of squares, rectangles, circles and precision printed screens are produced by the various manufacturers. Narrow, sharp-edged adhesive tapes come in various widths and they are ideal to make single and multiple slits. Diffraction gratings can be produced by photographic reduction of parallel-line screens. Cubic, rhombic and other crystal structures may be simulated by copying appropriate two-dimensional patterns. Tapestries, wallpapers and snowflake patterns are just a few additional ideas for use as aperture masters. To save film, 4 to 6 small aperture designs can be glued on one letter-size white paper and photographically reduced on one frame of 35-mm film.

A 35-mm single lens reflex is the most suitable camera for photographing the aperture masters. If the standard 50-mm lens does not focus close enough, a supplementary lens attachment or an extension tube (bellows extension) between the camera body and the lens will permit the lens to be focused at shorter distances.

Aperture masters up to 20x25 cm can be illuminated evenly by placing two No. 1 photofloods at an angle of 45-degrees to the lens axis as shown in Fig. 206.

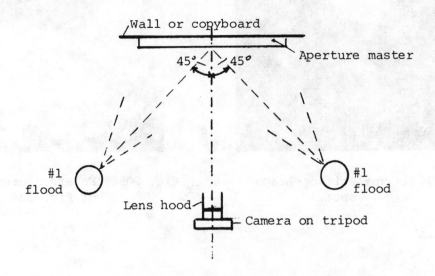

Fig. 206. Photographing the aperture master.

Reflector type bulbs (#1 photofloods) give good results without external reflectors but a lens hood is a necessity unless the copying lights are in deep reflectors. Room illumination should be turned off during exposure. To minimize flare the lens must be absolutely clean.

Kodak High Contrast Copy Film, Type 5069 (HC 135-36) is more satisfactory than Panatomic X for photographing aperture masters. Nominal resolving power exceeds 600 lines/mm. The Kodak Photomicrography Monochrome Film, SO-410 has excellent sensitivity and can be processed to a wide range of contrasts. This film has extremely fine grain, extremely high resolving power. The SO-410 can be processed in hand-processing equipment, such as spiral reels in small tanks. The film is sold preloaded in 36-exposure magazines.

A reflected-light meter reading should be taken from the gray (18% reflectance) side of the Kodak Neutral Test Card at the surface of the copy-board to obtain a reading for a trial exposure. Extra exposure will have to be calculated to compensate for a long bellows extension. If no Neutral Test Card or light meter is available, set the shutter for 1/60th second and make three exposures: one at f/4, one at f/5,6 and the last one at f/6,3. The two #1 reflector floods should be placed about 50 to 60 cm from the center of the aperture master to be photographed. The exposed film is developed in Kodak D-19 for 4 minutes at 20 C (68 F) with agitation at 30-second intervals. Rinse for 30 seconds with agitation in Kodak Stop Bath SB-1a and fix in Kodak Fixing Bath F-5 for 2-4 minutes. Wash in clear, running water for 10-15 minutes. Rinse one minute in Kodak Photo-Flo Solution after washing to minimize drying marks. Rinse, fix, and wash at 18-21 C (65-70 F). Dry in dust free area. Cut the film and mount the negative transparencies (aperture masks) in glassless diabinders, such as made by "gepe." (The exposure times and processing instructions are given in reference to the High Contrast Copy Film. This film gives excellent results and is much easier to obtain than the SO-410 type.)

For greatest sharpness, the copy camera and the copy board must be on solid support. Vibration from whatever source will impair sharpness, as will overexposure and/or overdevelopment.

A wide variety of dry-transfer stock patterns are manufactured by Artype, Prestype, Formatt, Visi-Graphics, Letraset, Chartpak (Avery) and Normatype. The black tape made by Chartpak (#352) is a 0.9 mm wide self-adhesive tape, excellent for making single, double and multiple slit masters.

References:

F.S. Harris, Jr., Appl Opt v3 p909 (1964)
Kodak Publication No. P-52, "Techniques of Microphotography" (3-1967)
V. Mallette, Phys Teach p52 (Jan '73)
R.C. Nicklin and J. Dinkins, Phys Teach p295 (May '74)
Kodak Publication No. P-304, "Kodak Photomicrography Monochrome Film, SO-410,
(2/73 rev.)
Kodak Professional Data Book, No. M-1, "Copying, 8th ed." (1974)

" " " " " " " " " " "

Simple Optical System for Fraunhofer Diffraction Experiments

When an aperture (or obstacle) is placed within the cross section of a beam of light and the transmitted light is observed on a screen, the resulting distribution of light is called a diffraction pattern. There are two basic forms of diffraction: Fresnel and Fraunhofer. If both source and observation screen are at a finite distance from the diffracting object, the pattern is called a Fresnel diffraction pattern (near-field diffraction). Under these conditions an image of the aperture (or obstacle) is projected onto the observation screen. The image is clearly recognizable despite some fringing around its periphery. If the observation screen is placed very far from the diffracting object, the pattern is called a Fraunhofer diffraction pattern (far-field diffraction). At a very great distance the projected pattern will bear little or no resemblance to the diffracting object. Moving the observation screen even further away will only change the size of the pattern but not its shape.

The near-field extends from the diffracting object to a distance of D^2/λ, where D is the diameter of the diffracting object and λ the wavelength, while the far-field extends from there to infinity. For example, for an aperture of 2.5 cm (1") diameter and He-Ne laser light, the far-field observation point has to be more than 1,600 meters or one mile!

The far-field conditions can be closely simulated by using a lens to produce a plane parallel (collimated) beam from a point source and another lens to bring the beam to a focus (Fig. 207).

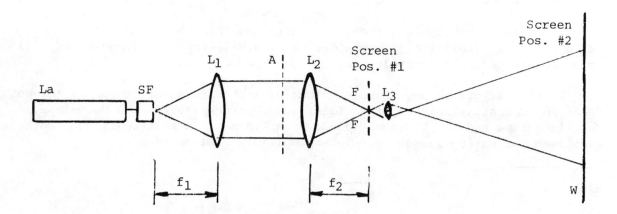

Fig. 207. Optical diffractometer.

The spatially filtered light that emerges from the pinhole is collimated by lens L_1. A second lens L_2 brings the expanded beam to a focus in its focal plane FF. If an object A is placed between the lenses, its Fraunhofer pattern is seen at FF (Screen position #1). A short focal length lens L_3 may be added to project an enlarged image of the pattern on the screen in position #2.

By using a large, well-corrected lens, L_2, the diffracted waves can be focused on a nearby screen without substantially changing the shape of the pattern. It is not even essential that the incident wave be planar so long as the optical path lengths for all rays, from the diffracting object to the viewing screen are essentially equal. A simple demonstration setup is shown in Fig. 208.

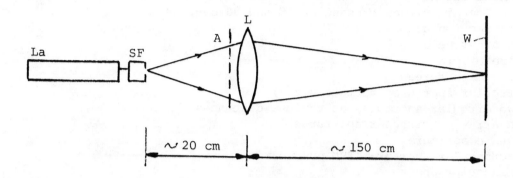

Fig. 208. Demonstration setup for Fraunhofer diffraction.

The spatial filter (SF) is Metrologic 60-618 (lens 36 mm f.l., and the pinhole about 30 μm diameter). L is a 167 mm f.l. uncorrected positive lens. The lens is adjusted on the optical bench to focus the image of the pinhole on the screen, W. The diffracting objects (apertures and obstacles) are inserted at A. The distance between A and the collecting lens L is immaterial. If the distance changes, the pattern does not change. With increasing distance the field of view will be vignetted, and the outer fringes cut off. With the given setup the diffracting objects used are about 5-7 mm in diameter. Note that, unlike the Fresnel pattern, the Fraunhofer pattern seen on the observation screen bears little resemblance to the geometry of the diffracting object. To observe a Fresnel pattern of the diffracting object, remove the lens from the optical bench while keeping A in place.

To obtain a large field of view, a large aperture lens L would be required. It is then more convenient and less expensive to replace the lens L by a concave spherical mirror. One then avoids the effects caused by striations in the lens itself. The mirror, M (for example: Edmund No, 40,913, 25" f.l.), must be used with oblique illumination in order to avoid double passage through the diffraction aperture (Fig. 209).

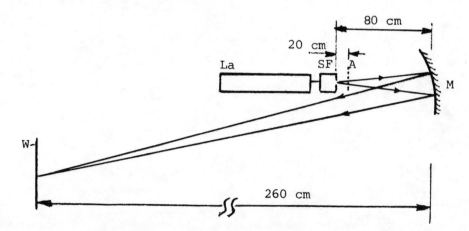

Fig. 209. Mirror setup for Fraunhofer diffraction demonstration.

189

Some suggested diffraction apertures (obstacles):

    Single slit
    Double slit
    Multiple slits
    Ronchi ruling
    Two Ronchi rulings crossed at 45 and 90 degrees
    Diffraction grating
    Straight edge
    Wire or pin
    Wire mesh screen
    Circular aperture (various sizes)
    Two circular apertures at various distances
    An array of circular apertures
    Square aperture
    Rectangular aperture (IBM punchcard)
    Triangular aperture
    Pentagonal aperture
    Hexagonal aperture

References:

D. Dutton, M.P. Givens, and R.E. Hopkins, Am J Phys v32 p355 (1964)
E.V. Palmer, and J.F. Verril, Contemp Phys v9 p257 (1968)
S. Berko, Y.G. Lee, F. Wright, and J. Rosenfeld, Am J Phys v38 p348 (1970)
E. Hecht, Am J Phys v40 p571 (1972)
G.R. Graham, Phys Educ v7 p352 (1972)
R. Bergsten, Am J Phys v42 p91 (1974)
M.I. Darby, and N. Morton, Phys Educ v9 p361 (1974)
M.J. Maloney, and W. Meeks, Am J Phys v42 p696 (1974)
A. Leitner, Am J Phys v43 p59 (1975)
H. Weichel, and L.S. Pedrotti, El-opt Syst Design p14 (Dec '75)
P.B. Pipes, and T.F. Dutton, Am J Phys v44 p399 (Apr '76)

"""""""""""

Fraunhofer diffraction at a circular aperture is of great importance as most lenses and stops used are circular in shape.

The diffraction pattern formed behind a circular aperture consists of a bright central disc surrounded by dim concentric annular rings; the rings are separated by dark bands.

The phenomenon that point images could not be formed of point objects even in an ideal, aberration free optical system, was investigated by the English astronomer Sir George Airy in 1835. The bright central portion of the intensity pattern is referred to as "Airy's disc."

Since the resolving power of a uniformly illuminated lens is limited by the size of the Airy disc, a lens called 'diffraction limited' is the best that can be achieved.

A simple setup for the observation of the Airy disc is shown in Fig. 210.

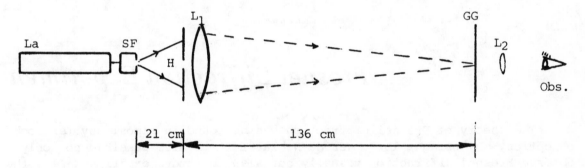

Fig. 210. Setup for the observation of the Airy's disc.

H is a commercial pinhole or a hole about 0.1 to 0.2 mm in diameter made in a household aluminum foil. Make several holes with different size needles to obtain 6-7 diffraction rings on the ground glass screen (GG). If the pinhole is not exactly circular, the fringes will not be exactly circular.

The numerical values (distances) shown in Fig. 210 are approximate for $L_1 = 167$ mm focal length. The pinhole of SF is focused on the ground glass screen (GG). The aperture H is very close to lens L_1. Lens L_2 is a magnifying lens. The screen (GG) may be replaced with a TV camera (vidicon tube) without the camera objective to demonstrate the Airy disc to a larger group of people. A neutral density filter may have to be placed at the output of the laser to assure that the photocathode of the vidicon tube is not saturated by the intense central disc.

_____

References:

D. Keeling, Photog Applic Sci Technol & Medicine p20 (Mch '74)
P.M. Rinard, Am J Phys v44 p70 (Jan '76)

Fig. 211. Fraunhofer diffraction pattern of a circular aperture.

" " " " " " " " " " " "

Fresnel Diffraction Experiments

The theory of Fresnel diffraction can be found in almost any textbook
on optics. By using a low-power He-Ne laser it is also possible not only to
observe Fresnel diffraction visually but also to study, experimentally, the
intensity distribution in the Fresnel diffraction pattern of various apertures
and obstacles.

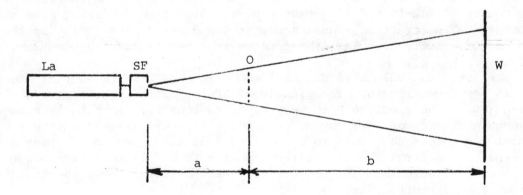

Fig. 212. Demonstration of Fresnel diffraction.

The pinhole of the lens-pinhole spatial filter (SF) acts as a point
source for Fresnel diffraction.   The diffracting aperture or mask is placed
at O, near the pinhole, about 5 meters (15 ft.) from the screen, W. It is
fascinating to slide the diffracting object between the pinhole and the screen

along the axis of the diverging laser beam and observe the remarkable changes in the diffraction pattern.

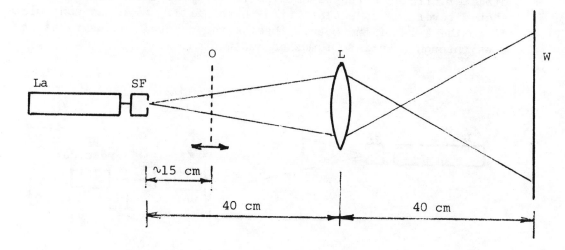

Fig. 213. Arrangement to project Fresnel patterns.

Figure 213 depicts an arrangement for demonstrating Fresnel diffraction patterns to a group of observers. The beam from the laser (La) is spatially filtered (SF). The diffracting object (O) is placed in the divergent beam. A projector lens (L = 3-4 cm f.l.) is placed about 40 cm downstream from the spatial filter. The diffraction pattern is projected on the screen (W) about 40 cm beyond the lens. The magnification can be changed by moving the diffracting object (O = aperture or obstacle) toward or away from the spatial filter.

Some suggested demonstrations:

(a) Single slit Fresnel diffraction: use a variable-width slit or photographic slides of slits with increasing slit width.

(b) Circular apertures: use an iris diaphragm to vary the aperture diameter or photographic slides of circular apertures with increasing aperture diameter.

(c) Circular obstacle: the head of a pin or various size ball-bearings glued to a glass plate.

(d) Straight edge diffraction: a scalpel blade or a razor blade. A single wire or needle. Two straight wires or needles parallel and close together. Place a plano convex lens or convex mirror on an optical table. Raise the table until the convex surface is introduced into the beam path.

(e) Rectangular aperture: a hole in an IBM punch-card is convenient.

(f) Double slit: Photo slide of double slits with various slit spacings.

(g) Multiple slits: Ronchi rulings.

(h) Two-dimensional gratings: Two crossed Ronchi rulings.

(i) Fresnel zone plates: Make your own zone plate by photographing the
illustration (elsewhere) in this book. Use Kodak High Contrast
Copy Film and reduce the zone plate illustration to cover a 5 mm
diameter circle on the 35-mm film frame. The divergent laser beam
should cover the zone plate area on the film. Move the zone plate
along the axis of the beam. Observe the central intensity as it
goes through a series of maxima and minima.

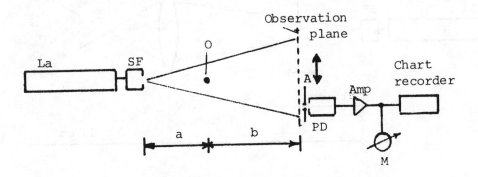

Fig. 214. Fresnel diffraction by a thin wire.

Figure 214 shows the arrangement to verify the law of intensity in the
Fresnel diffraction of a thin wire. The laser beam is filtered with SF.
At distance a (10-15 cm from the pinhole) place wire O, (diameter 0.1 mm).
The Metrologic 60-255 photo-detector PD, is placed on a home-made motor-driven
cross slide at a distance b (3 meters from the wire). A 0.2 mm aperture is
mounted in front of the photo-detector to permit intensity measurements of a
small portion of the pattern. The amplified output of the photo-detector is
fed to an ammeter or a chart recorder to plot accurate measurements of the
intensity versus accurate measurements of position - point-by-point or
continuously.

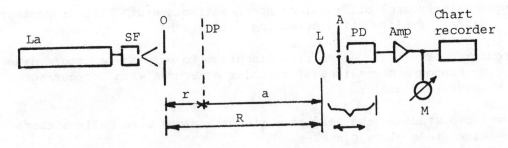

Fig. 215. Fresnel diffraction by a round aperture.

Figure 215 shows a sketch of the experimental setup for studying the
distribution of the illumination along the axis of a round aperture. This
exercise illustrates the application of the Fresnel-zone method for the study
of diffraction problems. A round aperture O of about 1.0 mm diameter is
centered on the axis of the diverging laser beam. The measuring unit consists
of a lens, L (30-50 mm f.l.), about 300 mm in front of the photo-detector PD,
and a small field diaphragm A which allows only the center of the magnified
image of the central diffraction spot to reach PD. The lens, diaphragm, and
photo-detector form a unit and are moved together during the course of the
measurements. The lens L forms a magnified image of the diffraction pattern

produced in the plane DP conjugate to the lens. The magnification of the diffraction pattern is the same for any plane of DP. To plot the distribution of illumination along the axis of a round aperture one measures the amplified photo-detector output and the distance R from the lens L to aperture O to determine the distance r from the aperture to the investigated plane DP (r = R-a).

A simple method for real-time quantitative study of Fresnel diffraction by a slit is shown in Fig. 216.

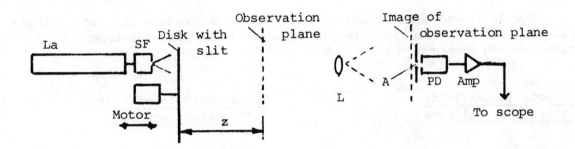

Fig. 216. Real-time demonstration of Fresnel single slit diffraction pattern on an oscilloscope.

The laser beam is filtered and made divergent with SF. A short focal length convex lens L is located some distance from the laser. A photo-detector PD is centered in the expanded beam. A small circular disk with a single radial slit (width about 0.5 mm) cut from the outer rim to halfway to the center of the disk is mounted on the shaft of a small motor. When the rotating disk is inserted into the laser beam the slit is swept through the beam and the image of the diffraction pattern is swept past another slit or pinhole A (0.05 mm dia.) in front of the photo-detector, PD. The amplified signal is fed to an oscilloscope.

While in Fig. 214 the thin wire was in fixed position and the observation plane was scanned by moving PD, In Fig. 216 the observation point is fixed and the slit is moved by rotating the disk with a motor. By moving the rotating disk closer or further away from the lens the changing diffraction pattern can be continuously monitored on the oscilloscope. Thus, the continuous transition from Fresnel to Fraunhofer diffraction can be viewed directly.

References:

J.D. Barnett, and F.S. Harris, J Opt Soc Am v52 p637 (1962)
R.E. Haskell, Am J Phys v38 p1039 (1970)
A.L. Moen, and D.L. Vander Meulen, Am J Phys v38 p1095 (1970)
L.I. Vidro, Yu.P. Basharov, and A.E. Kudryashov, Sov Phys-Uspekhi v13 p826 (1971)
R. Boyer, and E. Fortin, Am J Phys v40 p74 (1972)
T.W. Eaton, and D. Wiseman, Phys Educ v11 p292 (1976)

" " " " " " " " " "

Diffraction Patterns

The diffraction patterns shown below were recorded on Kodak Panatomic-X film. A 0.9 mW laser (Metrologic ML-968) was the light source. The Fresnel patterns were photographed with the setup shown in Fig. 213. The arrangement shown in Fig. 208 was used to photograph the Fraunhofer patterns. The film in the 35-mm single lens reflex camera took the place of the screen, W. Exposure times ranged from 1/60 to 1/250 second.

The diffracting masks and apertures used to produce the diffraction patterns are shown in Fig. 218. Mask/aperture numbers correspond to diffraction pattern numbers.

Fresnel (near-field)
diffraction patterns:

Fraunhofer (far-field)
diffraction patterns:

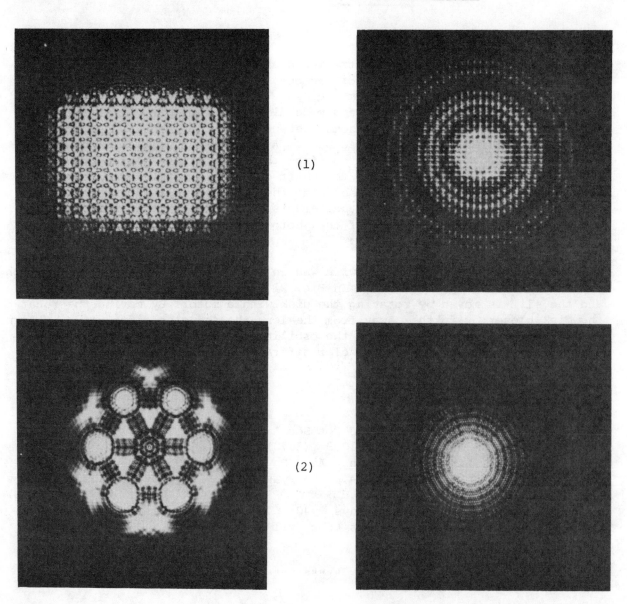

(1)

(2)

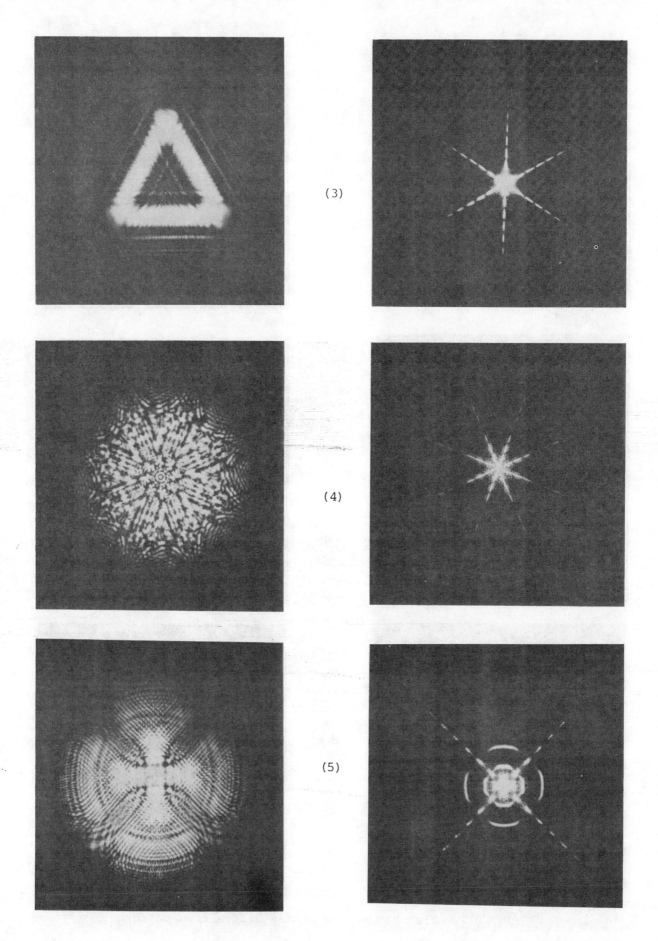

(3)

(4)

(5)

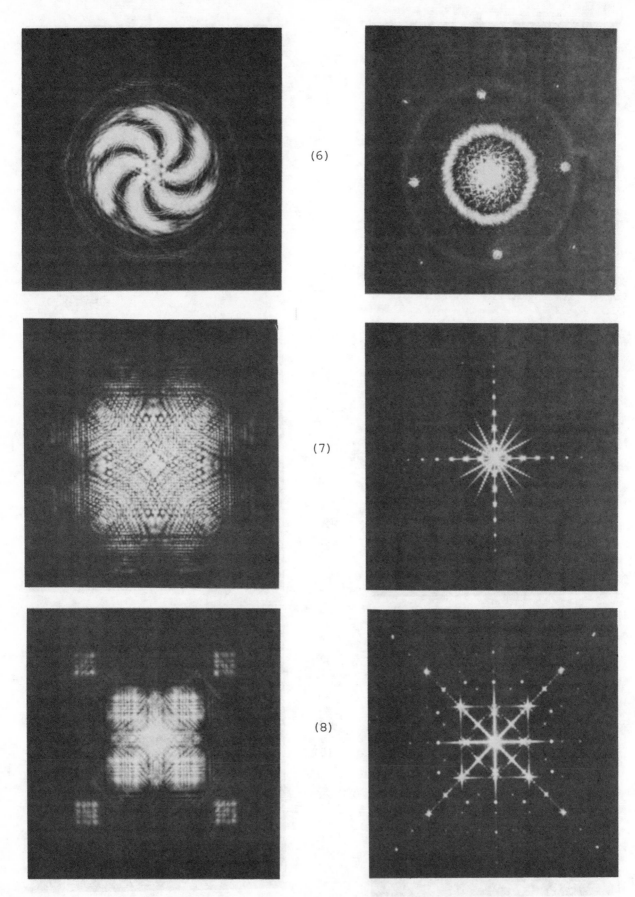

(6)

(7)

(8)

Fig. 217.   Fresnel and Fraunhofer diffraction patterns.

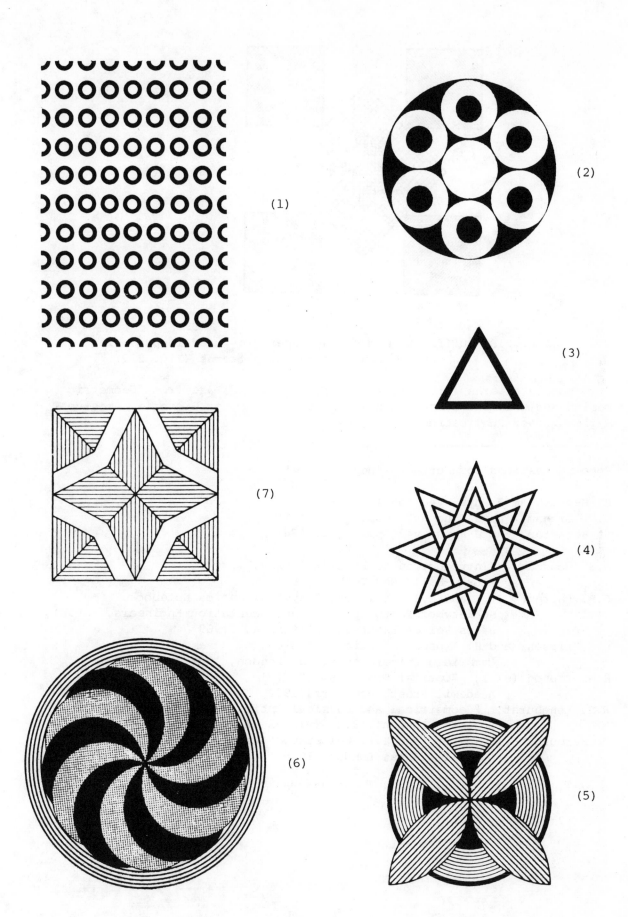

(1)

(2)

(3)

(7)

(4)

(6)

(5)

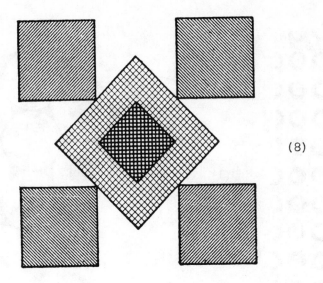

(8)

Fig. 218. Diffraction masks and apertures used in producing
Fresnel and Fraunhofer patterns (Fig. 217).

Note: Patterns (2), (4), (5), (6), (7) and (8) are from "Geometric
Design and Ornament," by Edmund V. Gillon, Jr., Dover Pictorial Archive
Series, Dover Publications, Inc., New York.

Some Suggestions for Further Study:

E. Hecht, Am J Phys v40 p571 (Apr '72)
G.R. Graham, Phys Educ v7 p352 (Jly '72)
R. Bergsten, Am J Phys v42 p91 (Feb '74)

J.W. Goodman, "Introduction to Fourier Optics,"
                McGraw-Hill, New York, 1968
G.B. Parrent, Jr., and B.J. Thompson, "Physical Optics Notebook,"
                Society of Photo-optical Instrumentation Engineers, (SPIE),
                Palos Verdes Estates, California, 1969
S.G. Lipson, and H. Lipson, "Optical Physics,"
                Cambridge University Press, London, 1969 (pp470-473)
H.S. Lipson (ed.), "Optical Transforms,"
                Academic Press, New York, 1972
R.S. Longhurst, "Geometrical and Physical Optics,(3rd edition),"
                Longman, London, 1973 (Plates II and III)
G. Harburn, C.A. Taylor, and T.R. Welberry, "Atlas of Optical Transforms,"
                G. Bell & Sons, London, 1975

" "" "" "" "" "" "

Measuring the Wavelength of Light with a Ruler

Diffraction of light may be observed in common everyday objects. One such object is an LP phonograph record. The grooves act as a reflecting grating and good diffraction spectra can be observed on a nearby wall if a laser beam is bounced off the LP record at an oblique angle.

A.L. Schawlow described a lecture demonstration in which a laser beam is diffracted at grazing incidence by the rulings of a steel (or plastic) meter scale. The wavelength of the light is obtained by measuring the pattern spacings and the distance from the ruler to the screen. The arrangement is shown in Fig. 219.

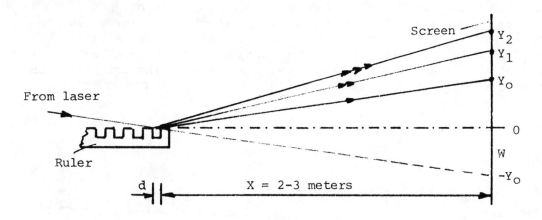

Fig. 219. Schawlow's experiment with a ruler.

All information required to measure λ comes from the spacings of the spots on the screen and the distance from screen to ruler. The wavelength of the laser light is determined from the formula:

$$\lambda = \frac{d}{2n} \frac{(Y_n^2 - Y_o^2)}{x^2}$$

where d is the spacing between rulings; n is in integer (the diffraction order); Y_n and Y_o are measured from 0 along the projection screen. The intersection of the plane of the grating with the screen (0) lies halfway between the spots of the direct beam ($-Y_o$) and the zero order diffracted beam which is specularly reflected (Y_o); x is the distance between the ruler and the screen.

References:

A.L. Schawlow, Am J Phys v33 p922 (Nov '65)
G.L. Rogers, Phys Educ v11 p346 (Jly '76)

" " " " " " " " " "

Measuring the Diameter of
a Hair by Diffraction

It is often necessary to measure the diameter of fine wire, hair, yarn or filament. The diffraction of laser light has been used in industry to measure the thickness of fine yarn or wire during the fabrication process.

A laser beam is positioned incident on a filament, such as a wire, to produce a diffraction pattern. By measuring the spacing between the light (or dark) areas of that diffraction pattern the diameter of the filament may be obtained by solving an equation in which the filament diameter is the only unknown.

As illustrated in Fig. 220, the laser beam is incident upon a double slit (S_1 and S_2) spaced apart by a distance d. An observation screen, W, is located at a distance D from the plane of the double slit. A diffraction pattern, including a plurality of alternating bright and dark regions is produced on the screen. The distance between the adjacent bright regions (and the dark regions) formed on the screen is designated Δy.

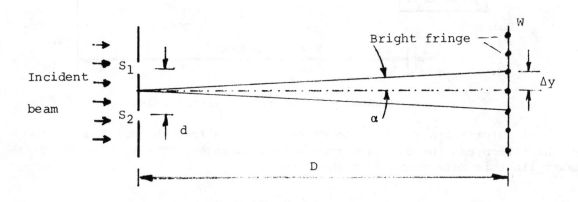

Fig. 220

From Bragg's law of diffraction it is known that $d \sin \alpha = m\lambda$, where $\underline{m}$ is an integer, λ is the wavelength of the light and α is the angle shown in Fig. 220. Assuming that Δy is much smaller than D, and $\tan \alpha$ is approx. equal to $\sin \alpha$, we obtain $\lambda = D \sin \alpha$. Substituting for $\sin \alpha$ we then obtain $\lambda = d\Delta y/mD$ or $d = \lambda mD/\Delta y$. Thus, the distance of $\underline{d}$ between the slits S_1 and S_2 can be determined by measuring Δy, since λ, m, and D are known.

As shown in Fig. 221, the slits are replaced by the diameter of a wire on which the collimated laser beam is incident to produce the diffraction pattern on screen W.

References:

H.A. Kruegle, US Pat.No. 3,709,610 (9 Jan '73)
S. George, and M. Guarino, Phys Educ v8 p392 (Sep '73)
S.M. Curry, and A.L. Schawlow, Am J Phys v42 p412 (May '74)

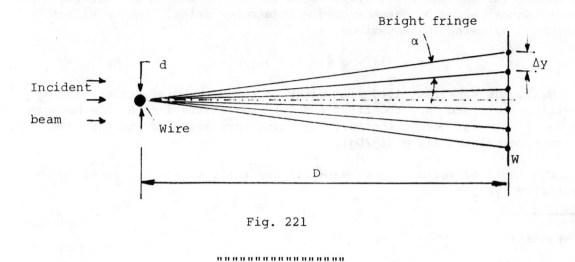

Fig. 221

" " " " " " " " " " " " " " " "

Diffraction Refractometer

The wavelength of light can be measured easily by use of the diffraction grating. A laser, a fish tank and an inexpensive replica grating is all that is needed to calculate the wavelength of the laser light in air and in water. From the ratio of the two wavelengths the index of refraction of water can be obtained.

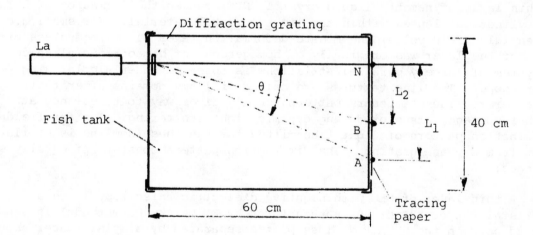

Fig. 222. Experimental arrangement of the refractometer (top view).

The laser beam enters near one end of an empty fish tank. The tank is adjusted so that the reflections of the beam are bounced back from the glass surfaces of the tank to the laser source. An inexpensive replica grating (from Edmund Scientific Co.) is taped on the inside of the fish tank in the path of the laser beam. These gratings have 13,400 lines per inch (5,276 lines per cm). A tracing paper is taped on the outside wall of the tank opposite the laser. With air in the fish tank, the laser is turned on and the normal (N) and the first order image (A) are marked on the tracing paper. The distances L_1 and D_1 are measured with a meter stick. The wavelength is determined by using the equation:

$$\lambda_1 = d \sin \theta = dL_1/D_1$$

Next, the tank is filled with water to cover the diffraction grating. It will be observed that the first order fringe shifts from A to B because the wavelength had become shorter. L_2 and D_2 are measured and the wavelength in water calculated ($\lambda_2 = dL_2/D_2$).

The index of refraction of water is the ratio of the two wavelengths: n(water) = λ_1/λ_2.

References:

M.L. Clark, Phys Teach v2 p85 (Feb '64)
A.B. Nafarrate, IBM Tech Discl Bull v13 p121 (Jne '70)
R.F. Kotheimer, Phys Teach v12 p307 (May '74)
A.T. Magill, Phys Teach v13 p555 (Dec '75)

" " " " " " " " " "

Electrically Controllable Diffraction Grating

R. Williams discovered in 1963 that application of an electric field to a thin layer of nematic liquid crystal (NLC) causes the formation of a lattice of cylindrical lenses within the liquid crystal material. For small a.c. potentials (5-10 volts) the minute deformation takes place regardless of whether the layer was originally of homogeneous or homeotropic alignment. A system of largely parallel striations is observed under a polarizing microscope. The line textures are called Williams domains after their discoverer. The stationary pattern usually gives way to turbulence at higher voltages, thus producing dynamic light scattering. The far-field diffraction pattern of light transmitted through these domains is similar to that from a phase grating. The diffraction pattern consists of equally spaced fringes.

A thin layer of NLC with negative dielectric anisotropy (such as Licristal #5A or #7A or #9A manufactured by E. Merck, Darmstandt) is sandwiched between two NESATRON Glass plates separated by a Mylar spacer about 15-50 μm thick. Orientation is achieved by the method of rubbing the conductive surfaces with a clean cloth or tissue paper. As the cell is assembled the rubbing directions are aligned parallel.

Figure 223 shows the arrangement to demonstrate light diffraction by a NLC phase grating. A He-Ne laser beam is normally incident on the liquid crystal cell. The polarization direction of the incident laser beam is parallel to the rubbing direction. A far-field diffraction pattern can be observed on the screen W when the cell is properly activated. The optical effects are much more reproducible when audio frequency a.c. fields are

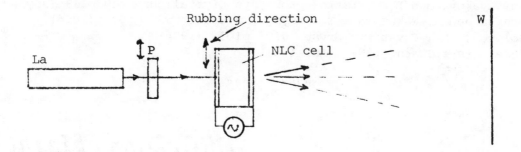

Fig. 223. A schematic diagram for observing light diffraction from
a liquid crystal phase grating (Williams domains).

employed. Very little frequency dependency is observable from 50 to 500 Hz.
The voltage for the first observation of the domain pattern is insensitive
to sample thickness. The domain pattern is quite stable as the voltage is
increased above threshold. At roughly twice the critical voltage the lines
begin to move and at still higher voltages the cell begins to operate in the
dynamic scattering mode. The spacing between the domain lines (The period
of the phase grating) is approximately the same as the thickness of the NLC
layer.

It is hard to change the period of the grating if it is formed of
Williams domains. It is possible, however, to produce tunable optical
diffraction gratings by using very thin (3-10 μm) layers of highly purified
NLCs (resistivity more than 10^{10} ohm cm), such as Licristal #4 or #5
(E. Merck, Darmstadt). Domain formation begins at 6-10 volts. If the NLC
layer is about 6 μm thick (1/4 mil), the domain width steadily becomes
smaller with increasing field strength. As the domain width decreases, the
total number of domains will increase. The effect of electrically controllable
domains in NLCs was first described by Greubel and Wolff. They reported that
using highly purified NLCs in ultra-thin layers the inverse domain width is
proportional to the applied electric field and that dynamic scattering does
not occur up to the highest practicable field strength. The domains are best
visible when the incoming laser light is polarized perpendicular to the long
axes of the domains, and vanish for parallel polarized light. In rubbed
cells (with and without a surfactant), the long axes of the domains are all
perpendicular to the direction of rubbing.

By illuminating the cell with a parallel (collimated) He-Ne laser beam
normal to the glass plates, one observes diffraction spots on the screen W
with suitable polarization of the laser beam. With a.c. excitation the
domain width oscillates with twice the frequency of the sine-wave source.
The described effect may find application in light modulators, deflectors and
controllable optical filters.

References:

R. Williams, J Chem Phys v39 p384   (15 Jly '63)
W. Greubel, and U. Wolff, Appl Phys Lett v19 p213   (1 Oct '71)
T.O. Carroll, J Appl Phys v43 p767   (Mch '72)
C.L. Hedman, Jr., et al, U.S. Pat. No. 3,758,195 (11 Sep '73)

R.A. Kashnow, and J.E. Bigelow, Appl Opt v12 p2302 (Oct '73)
I.G. Chistyakov, and L.K. Vistin, Sov Phys Crystallogr v19 p119 (Jly/Aug '74)
H. Tsuchija and K. Nakamura, Mol Cryst Liq Cryst v29 p89 (1974)
P.H. Bolomey, Mol Cryst Liq Cryst v29 p103 (1974)
P.A. Penz, Phys Teach p199 (Apr '75)

"""""""""""

Diffraction of Light by Ultrasonic Waves

Among the techniques for controlling the output radiation of laser light sources, the interaction of ultrasonic and optical beams is proving to be one of the most versatile and useful.

Ultrasonic waves are sound waves whose frequencies are higher than those normally audible to the human ear.

In 1921 L. Brillouin predicted that a liquid traversed by compression waves of short wavelengths, when irradiated by visible light, would give rise to a diffraction phenomenon similar to that caused by a ruled grating.

It was a decade after Brillouin's prediction that P. Debye and F.W. Sears observed the diffraction of light by ultrasonic waves.

When ultrasonic waves are propagated in a liquid medium, compressions and rarefactions of the medium follow one another through the medium at the velocity of the sonic wave. The distance between successive peaks of compression is the wavelength λ_S of the sonic wave. The variations in compression give rise to variations in the density of the liquid. The density gradient - and accordingly the variation of the refractive index - acts as a phase grating which will diffract a light beam incident on the liquid normal to the propagation direction of the sonic wave. As an example, consider an ultrasonic wave of frequency ν = 2 MHz propagated in water. Since $v_S = \nu\lambda_S$ the separation of regions of compression and rarefaction = $=\lambda_S/2 = v_S/2\nu$ equals approximately $(1.5 \times 10^5)/(2 \times 2 \times 10^6)$ = 0.0375 cm.

The distance between the various order diffraction is dependent on the wavelength of the light λ_L, the ultrasonic wavelength λ_S, and the angle of deviation θ_n for the n-th order spectrum is given by

$$\sin \theta_n = n\lambda_L/\lambda_S = n\lambda_L\nu/v_S \quad (n = 0, 1, 2, 3, \ldots)$$

where ν is the sound frequency and v_S its velocity. Diffraction of light by a liquid carrying a sound wave enables a direct measurement of the acoustic wavelength in the liquid and hence the acoustic velocity.

A) Setup for demonstrating sound-diffracted light patterns

The direct laser beam is incident at right angles to the ultrasonic wave (Fig. 224). In addition to the liquid, the 'ultrasonic cell' contains a piezoelectric crystal Q in a suitable holder. The crystal is excited at a

frequency of 5-10 MHz by a power oscillator. After the equipment is carefully
aligned so that normal or near-normal incidence of the laser beam and sound
column is achieved, the power oscillator is adjusted until a resonant frequency
of the crystal is obtained, at which point the diffraction pattern appears on
the screen, W. The screen is placed about 2 meters from the ultrasonic cell.
The reflector R is made of brass.

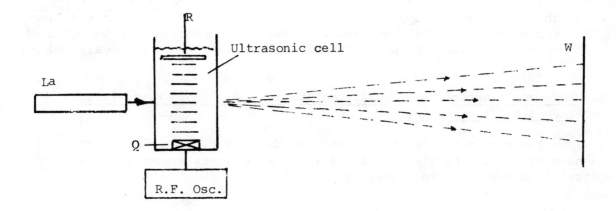

Fig. 224. Demonstrating light-sound interaction.

B) The Debye-Sears method for observing the diffraction of light by ultrasonic
waves in a liquid, and measuring the velocity of sound in various liquids.

The method is shown in Fig. 225. Divergent light from the lens-pinhole
spatial filter is collimated by lens L_1 and is passed through the ultrasonic
cell at right angles to the direction of propagation of the sound wave.
Lens L_2 forms the image of the pinhole at its focal plane on screen W. A
piezoelectric transducer Q is driven by an r.-f. oscillator and serves as the
sound generator. The metallic reflector R gives rise to the standing sound
wave.

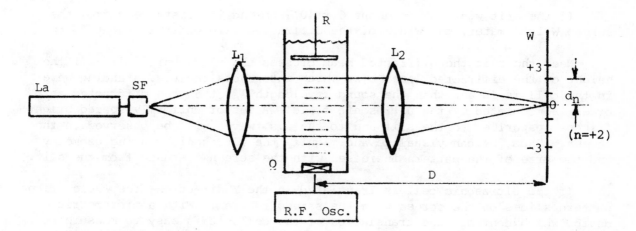

Fig. 225. Measuring the velocity of sound in liquids (Debye-Sears).

In order to produce the desired diffraction effects, the width of the
light beam must be larger than the wavelength of the sound. With zero voltage

applied to the u.s. transducer, all the light is undiffracted and falls in
the zero order at the center. On increasing the r.-f. voltage, more and more
light is diffracted, more orders appear, and less light falls in the location
of the zero order.

If D is the distance between the axis of the sound wave and the screen W,
and d_n is the distance between the zero order and the n-th order pinhole image
measured in the plane of W, we may write, assuming $D \gg d_n$, $v_s \approx (n\lambda_L \nu D)/d_n$.

Note: The value of D is approximately equal to the focal length of the long-
    focus lens L_2. The frequency ν of the r.-f. generator can be measured
    with an oscilloscope or preferably with an electronic frequency counter.

C) <u>Hiedemann's method for visualizing standing ultrasonic waves in the
    Fresnel diffraction region.</u>

The laser beam is limited in width by the slit F. The cross section of
the beam diffracted by the slit is observed on a screen W normal to the
optical axis x-x (see Fig. 226).

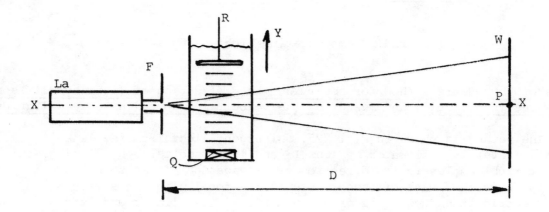

Fig. 226. Hiedemann's method.

If the slit width ΔF is about 6×10^{-3} mm and its distance D from the
screen W is 1 meter, the width of the diffracted zero order is about 20 cm.

Near the slit the diffracted beam crosses the ultrasonic cell. If the
height of the diffracted laser beam which intersects the u.s. standing wave
in the cell is larger than the sound wavelength λ_s, then the diffracted zero
order is modulated by the ultrasonic field. A set of uniformly spaced bright
fringes, paparllel to the slit and to the u.s. waves will be observed on the
screen. Thus, we can visualize and magnify the ultrasonic grating geometry
and the size of the pattern increases with the screen distance from the slit.

If the ultrasonic cell is translated in the Y direction, the whole fringe
pattern slides on the screen W in the same direction. With a micrometric
device which controls the translation of the cell, it is easy to measure
the ΔY displacement which is needed for the sliding of N fringes through the
intersection point P of the optical axis with the screen. A photoelectric
counter can be used at P for the automatic counting of the fringes. $\lambda_s/2$ is
given by $\Delta Y/N$. In order to have an absolute measurement of the ultrasonic
velocity, it is necessary to know the precise value of the frequency ν also.

D) <u>Projecting ultrasonic wavefront contours onto a screen using Hiedemann's method of secondary interferences due to the superposition of diffraction spectra.</u>

This method affords the direct visualization of the ultrasonic phase grating itself rather than the Fraunhofer diffraction pattern produced. The method also enables one to examine the wavefront contours over a broad cross section of the liquid column.

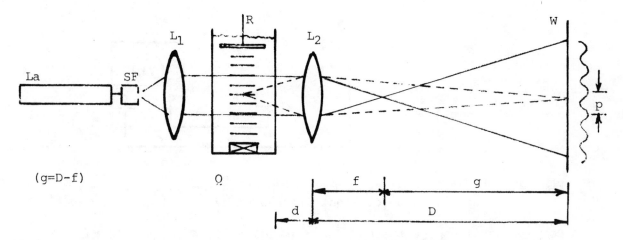

Fig. 227. Direct visualization of the u.s. grating itself.

The expanded laser beam is sent through the ultrasonic cell in which standing u.s. waves are generated in a liquid (see Fig. 227). The emergent light passes through or near the focal point of lens L_2 (about 5 cm f.l.) and then cast a 'shadow' of the sound waves on the screen W. The distance $g=(D-f)$ is approximately one meter. The distance d must be carefully adjusted to obtain the sharpest image on the screen because the degree of sharpness of the fringes varies periodically with the distance d. Try different values for d between 10 and 75 cm.

To determine the sound velocity, we use the equation: $v_s = 2\nu fp/g$, where p is the spatial period of the pattern observed on the screen, f is the focal length of L_2 the projection lens, and $g=D-f$ as shown in Fig. 227. The spacing p should be determined by measuring the span of typically 30-40 fringes with a ruler; and D can be measured with a meterstick.

E) <u>Displaying the intensity distribution of the light diffracted by traveling ultrasonic waves.</u>

Figure 228 shows the setup for displaying the intensity of the diffracted orders on an oscilloscope.

A plane front surface mirror M is attached to the shaft of a synchronous motor. As the mirror turns counterclockwise it reflects the light to detector PD_1 which triggers the oscilloscope and then to PD_2 which measures the light intensity which is then displayed on the oscilloscope. PD_1 should be so positioned as to provide the correct trigger delay. PD_1 and PD_2 are inexpensive photo-detectors. This rotating system can also measure the laser beam diameter and show its Gaussian intensity profile.

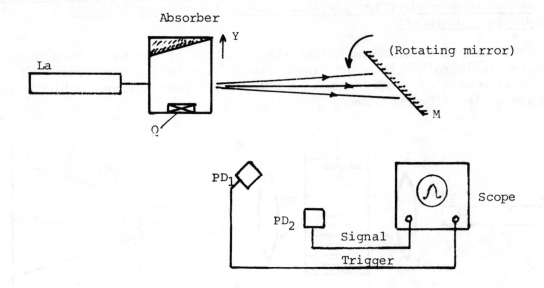

Fig. 228. Experimental system for displaying the intensity
distribution of light diffracted by traveling u.s. waves.

Compare oscilloscope traces, for example, at 2, 4, 6, and 10 MHz.
Observe the slight asymetry in the orders... this may be due to distortions
in the sound wave or due to the laser beam not being exactly parallel to the
sound wavefronts. Observe the effect of changing the r.-f. power applied
to the transducer. Observe the spatial distribution of acoustic energy in
the sound beam by translating the u.s. cell in the Y direction and obtain
information on the attenuation of the sound wave. Perform these exercises
with various liquids in the u.s. cell, such as, water, castor oil, and a
silicone fluid (Dow Corning 200).

F) Ultrasonic phase grating made visible by the dark-ground method.

     This technique shown in Fig. 229 is a variation of the Schlieren method.
Light from the pinhole of the spatial filter is made parallel by a condenser
lens L_1 and is passed through the u.s. cell, and is then refocused by a
second lens L_2 to a point image. A stop consisting of a spot of india ink
on a glass slide is placed at this focal point to exclude all direct light
from the pinhole to reach the viewing screen W. A third lens L_3 is placed
so as to collect the light in the diffracted images and form an image of the
cell on the screen. The screen will be dark until the transducer is excited,
when there will appear on the screen an image of the actual sound field in
the liquid. Traveling ultrasonic waves are used in this exercise and the sound-
absorber opposite the transducer Q is denoted by the letter A.

G) Ultrasonic light-modulating system.

     The system comprises an u.s. cell, an oscillator-modulator for controlling
the u.s. cell by the application of an amplitude-modulated r.-f. voltage, and
a signal source such as a microphone, phonograph or TV camera for applying a
signal voltage to the oscillator-modulator.

210

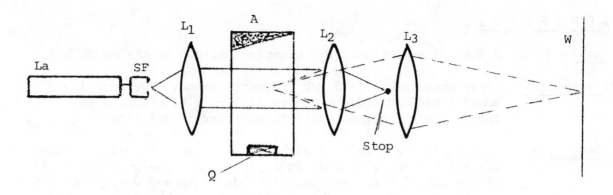

Fig. 229. Dark-field method for visualizing the u.s. phase grating.

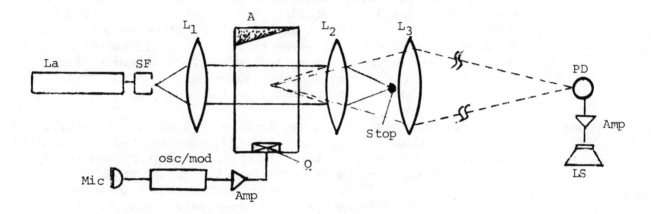

Fig. 230. U.s. modulator for voice-communication by means of
a laser beam.

While there is a similarity between the ultrasonic phase grating and the ruled grating, with the ultrasonic grating it is possible to change the proportion of light diffracted into a given order by merely changing the amplitude of the sound waves.

As shown in Fig. 230, an expanded and collimated laser beam passes through an u.s. cell and is then focused upon an intercepting obstacle. The traveling u.s. wave diffracts some of the incident light past the obstacle (stop) and thus generates an optical output whose intensity is a function of the acoustic power. (A sound absorber A is employed in the u.s. cell to prevent the formation of standing waves and spurious reflections).

With zero r.-f. voltage applied to the transducer Q, all the light is undiffracted and falls on the 'stop'. As the voltage and sound amplitude increase, diffraction images of the pinhole source are formed on either side of the stop. Light transmission increases rapidly up to about 90%. Thus, if the r.-f. voltage is turned up and down by hand, or if modulated electronically, the light passing by the stop is likewise modulated. A lens L_3 is placed so as to collect the diffracted light and focus it on a distant photo-detector PD. The amplified signal from PD is fed to a loudspeaker. The system can be used to provide simultaneous modulation of many speech channels on a beam of laser light. For a bandwidth of 4MHz, 200 speech channels, each 10kHz wide, could be accomodated.

APPARATUS:

Laser: 0.5 - 2.5 mW output. For example: Metrologic ML-968 or 969.

Electronics: Power oscillator: 1-15 MHz frequency range. For example:
                 General Radio Oscillator, Type 1211C; Viking Challenger
                 Transmitter; Lettine Radio Transmitter, Model 248.

U.s. cell: Rectangular tank made of aluminum or brass. Metallic housing
                 ensures screening of the liquid from the high-frequency field
                 of the oscillator. Microscope slides attached over ports cut
                 in the sides make suitable cell windows. They are cemented with
                 waterproof epoxy or Dow Corning RTV adhesive/sealant. Rectangular
                 glass tank may be used provided two opposite walls are of good
                 optical quality. Thick, adjustable brass plates are used as
                 reflectors for producing standing u.s. waves. If work is done
                 with traveling u.s. waves, the cell walls must be padded with an
                 absorber for compressional waves to prevent the formation of
                 standing waves and spurious reflections. Good absorbers are:
                 layers of ρc rubber; wire screens (50 mesh good up to 10 MHz -
                 three layers); fiberglass padding; cellulose sponge.

Transducer: (a) X-cut quartz crystal disc, 15-25 mm diameter, with metallized
                 surfaces, fundamental frequency 2-4 MHz, also excited at the third
                 and fifth harmonics; (b) Lead zirconate-titanate piezoceramic
                 disc, PZT-4 or PZT-5A Vernitron (Part No. 16035) or Gulton
                 HST-41 (Part No. 40225). Fine wire, such as #36 or thinner,
                 should be soldered to each side of the crystal using silver-
                 bearing solder. Low-power soldering iron (25-W or less) and a
                 minimum of solder should be used. Maximum efficiency of power
                 conversion is achieved when the crystals are 'air-backed' so that
                 most of the ultrasonic energy is radiated from the front of the
                 crystal. A porous material such as cork or foam rubber is
                 suitable for mounting the crystals with small amounts of adhesive.
                 Another method is to mount a 1-inch crystal over a 7/8" hole in
                 a polystyrene disc with Duco cement. If a metal vessel is used,
                 the crystal can be fastened over a slightly smaller diameter
                 hole with Eastman 910 Adhesive and/or Dow Corning RTV adhesive/
                 sealant.

Cell liquids: Water; Castor Oil; trichloroethylene; tetrachloroethylene;
                 Freon TF (nonflammable organic solvents); Dow Corning 200
                 silicone fluids (available in viscosities of 5-1,000 centistokes,
                 crystal clear).

References:

R.B. Barnes, and C.J. Burton, J Appl Phys v20 p286 (1949)
F. Porreca, J Acoust Soc Am v52 p427 (1972)
P. Kang, and F.C. Young, Am J Phys v40 p697 (1972)
D.T. Pierce, and R.L. Byer, Am J Phys v41 p314 (1973)
J.F. Neeson, and S. Austin, Am J Phys v43 p984 (1975)

continued...

Suggestions For Further Reading:

R. Adler, IEEE Spectrum p42 (May '67)
D.E. Flinchbaugh, Opt Spectra p49 (Sep '70)
K.N. Baranskii, and G.A. Sever, Sov Phys - Uspekhi v16 p161 (Jly-Aug '73)
M.B. Denton, and D.B. Swartz, Rev Sci Instrum v45 p81 (1974)
V.V. Maier, and V.E.-G. Khokhlovkin, Sov Phys-Uspekhi v16 p934 (May-Jne '74)
V.N. Mahajan, and J.D. Gaskill, Opt Engineering v14 p91 (Jan-Feb '75)
J. Lapierre, D. Phalippou, and S. Lowenthal, Appl Opt v14 p1549 (1975)
H.Z. Cummins, N. Knable, L. Gampel, and Y. Yeh, Appl Phys Lett v2 p62 (1963)

" " " " " " " " " " "

Self-imaging Without Lenses
(Fourier Imaging)

An optical image is the reproduction of an object formed by an optical system. An illuminated object is considered to be composed of an infinite number of adjacent points of light, each of which emits secondary radiation. Conventional optical systems employ a lens to convert the diverging radiation into a converging beam which intersects to form an image of the original object point. Other techniques usually employed would be dependent on interference techniques such as diffraction or self-imaging.

Self-imaging of periodic objects in coherent light is a well-known phenomenon. As far back as 1836 Talbot discovered that an image of a grating appears at integral multiples of the distance $2d^2/\lambda$ (where d is the grating period) when the grating is illuminated by a collimated monochromatic light beam. The phenomenon was interpreted theoretically in 1881 by Lord Rayleigh, who also used the effect to test and replicate diffraction gratings. The Talbot effect is frequently referred to in the literature as 'self-imaging' or 'Fourier-imaging'.

Using a Ronchi ruling, or two crossed Ronchi rulings, or a photographic transparency of a grid pattern (shown in Fig. 231) as a periodic object, the distance z_n of the n-th order image behind the object mask is given by $2n(d^2/\lambda)$. Sharp Fourier images of the periodic object form in these planes. Between the planes so defined, Fresnel images are formed which are not true images of the object.

Fig. 231. Cross-hatch pattern (a negative of this is used as target).

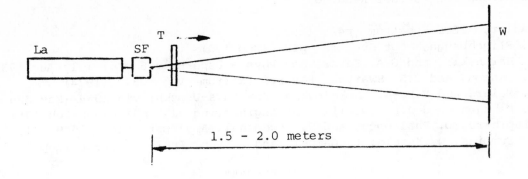

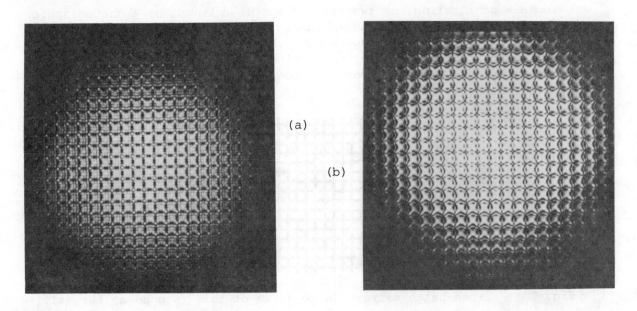

Fig. 232. Setup to demonstrate Fourier images.

Fourier images can be produced using either plane or spherical illuminating waves. Fig. 232 shows a simple demonstration setup with spherical wave illumination.

The laser beam is cleaned and made divergent with the spatial filter, SF. T is the photographic transparency of a periodic object. The cross-hatch shown in Fig. 231 was photographically reduced in size with a 35-mm single lens reflex camera. The grid pattern was pasted in the center of a large white cardboard and the image of the grid on the Kodak High Contrast Copy Film negative occupied an 8 mm x 8 mm area. An observation screen, W, is placed about 1.5 to 2.0 meters from the spatial filter. As the transparency is slowly moved from the vicinity of the spatial filter towards the screen, patterns of varying complexity are seen. These are Fresnel fringe patterns. At a certain distance from the pinhole source a sharp Fourier image of the grid is formed. Then the cycle is repeated. Figs 233(a) to 233(h) show some of the patterns observed. Figs. 233(c), (f) and (h) show the cross-hatch in the first, second, and third Fourier-imaging planes. It is clearly seen that the size of the Fourier images changes with propagation distance (because spherical wave illumination was used). The resolution of the self-images formed is of extremely good quality, obviating the need for any lens-type system.

(a)

(b)

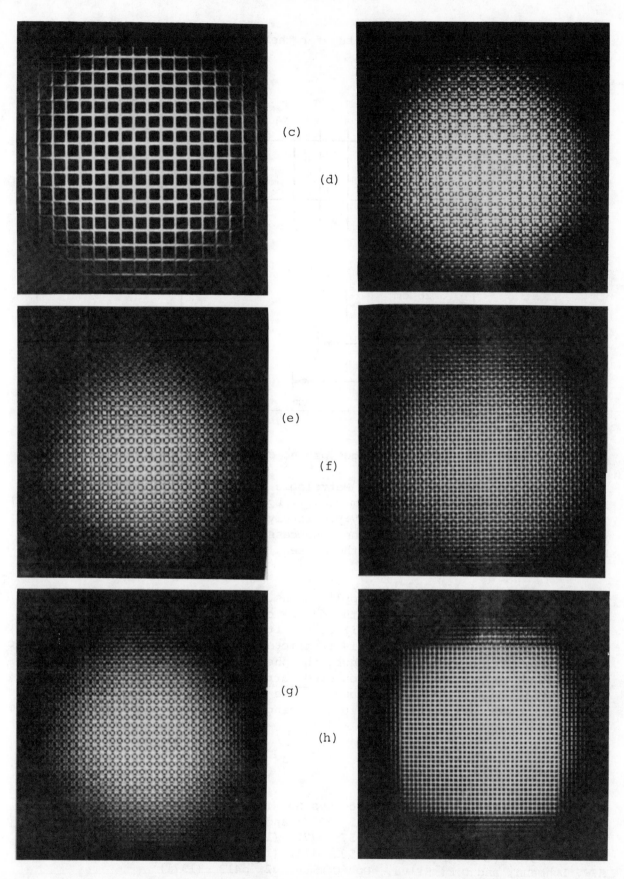

(c)

(d)

(e)

(f)

(g)

(h)

Fig. 233. Fresnel fringe patterns and Fourier images
produced by spherical illumination.

Figure 234 illustrates the setup for the production of Fourier images using plane illuminating waves.

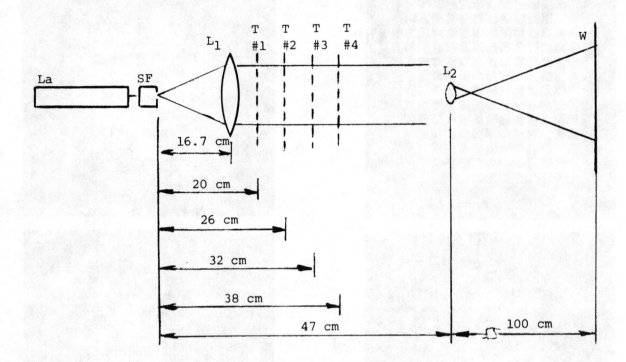

Fig. 234. Fourier images produced by collimated light.

The spatial filter, SF, is a Metrologic 60-618 used for this exercise. L_1 is the collimating lens (167 mm f.l.). L_2 is a projection lens (36 mm f.l.). T is the grid-pattern transparency previously described. Positions #1 to #4 indicate the approximate planes where successive Fourier images are formed. L_2 projects an enlarged image on the screen, W. All successive images are of the same size.

The Talbot effect has many applications. It is used in interferometry and also as a means to determine the degree of collimation. The effect can be used to align the light beam direction with respect to an optical bench. Since the Talbot imaging process is restricted to periodic objects, nonperiodic defects are not imaged at all. Thus, the phenomenon of self-imaging lends itself to the production of silicon diode array targets for vidicons (TV camera tubes) and the photomasks used in the manufacture of integrated circuits. Other applications include copying of diffraction gratings, multiple imaging and spatial filtering.

References:

J.M. Cowley, and A.F. Moodie, Proc Phys Soc (London) vB70 p486 p497
                              p505 (1957) and vB76 p378 (1960)
W.D. Montgomery, J Opt Soc Am v57 p772 (1967)
W.D. Montgomery, J Opt Soc Am v58 p1112 (1968)
A.W. Lohmann, and D.E. Silva, Opt Commun v2 p413 (1971)
H. Dammann, G. Groh, and M. Kock, Appl Opt v10 p81 (1971)
D.E. Silva, Appl Opt v10 p1980 (1971)

continued...

A.W. Lohmann, and D.E. Silva, Opt Commun v4 p326 (1972)
D.E. Silva, Appl Opt v11 p2613 (1972)
O. Bryngdahl, J Opt Soc Am v63 p416 (1973)
K. Patorski, S. Yokozeki, and T. Suzuki, Nouv Rev d'Opt v6 p25 (1975)
K. Patorski, S. Yokozeki, and T. Suzuki, Opt & Laser Technol p81 (Apr '75)
A. Kalestynski, Appl Opt v14 p2343 (1975)
B.J. Thompson, Appl Opt v15 p312 (1976)
S.A. Benton, and D.P. Merrill, Opt Engineering v15 p328 (Jly-Aug '76)

" " " " " " " " " "

Optical Simulation of the Electron Microscope

The sketch below illustrates the experimental setup which simulates an electron microscope (EM) imaging a two-dimensional lattice. The same setup can also be used to demonstrate x-ray diffraction effects.

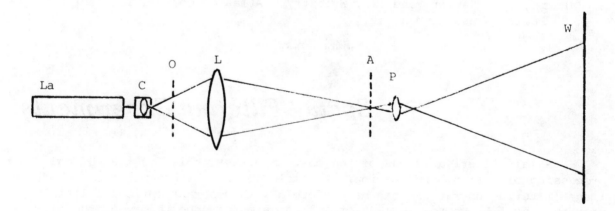

Fig. 235. Experimental setup.

A small He-Ne laser simulates the electron gun. A simple biconvex lens (C = 50-70 mm f.l.) simulates the condenser in the EM. A fine copper mesh, O (1,000 openings per inch), simulates a crystalline object in the EM. A good quality photographic lens (L = 50 mm f.l.) simulates the objective lens in the EM. (The back of this lens should face toward the object.) An aperture, A, cut out of cardboard, simulates the objective aperture in the EM. The objective of an optical microscope (P = 6 mm f.l.) simulates the projection lens in the EM. A white paper screen, W, simulates the fluorescent screen in the EM.

The optical components are assembled on an optical bench except the aperture, A, the position of which is adjusted frequently. With the above components, the distance between L and P should not be less than 600 mm.

The diffraction pattern (Fourier transform) of the object is easily observable in the back focal plane of L. Parts of the pattern can be easily blocked out (filtered) with various objective apertures, A. This allows the experimenter to demonstrate the features of Abbe's theory as well as the operation of the electron microscope.

This same setup can be used to demonstrate x-ray diffraction effects. The demonstration is based on the close analogy between x-ray diffraction by atoms and the optical diffraction of light by holes in a diffracting mask, O. The holes are arranged in a manner to simulate the positions of atoms in a projection of the unit cell of the crystal. The optical diffraction pattern produced by such a two-dimensional pattern of holes is the Fourier transform of the diffraction mask in exactly the same way that the x-ray diffraction pattern represents the Fourier transform of the electron distribution of the crystal. The aperture A should be removed from the setup.

References:

J.R. Meyer-Arendt, and J.K. Wood, Am J Phys v29 p341 (1961)
A.A. Balchin, and R.P.M. Dawson, Phys Educ v4 p58 (1969)
D.G. Ast, Am J Phys v39 p1164 (1971)
H. Lipson, "Optical Transforms," Academic Press, London-New York, 1972
R. Bergsten, Am J Phys v42 p91 (1974)
G. Harburn, C.A. Taylor, and T.R. Welberry, "Atlas of Optical Transforms,"
     G. Bell & Sons, London, 1975

" " " " " " " " " " " " "

Spatial Filtering Experiments

Spatial filtering is one of the most important tools for the optical processing of optical information. A simple but striking demonstration of the image-filtering operations can be performed with coherent optical filters as presented by A.B. Porter in 1906, to illustrate the Abbe theory of the microscope.

Figure 236 illustrates the conventional configuration of an optical system normally used for coherent optical data processing.

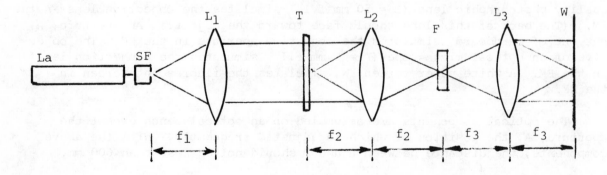

Fig. 236. Optical spatial filtering arrangement.

Light from a He-Ne laser is brought to a point focus by a short focal length lens. A pinhole is placed in the focal plane to eliminate laser beam 'noise.' This lens/pinhole spatial filter is shown in Fig. 236 as SF. Another lens, L_1, is placed a focal length, f_1, away from the plane of the pinhole to collimate the laser light. The input information, T, normally a photographic transparency, is then placed a focal length, f_2, away from the converging lens, L_2. This lens in its back focal plane, F, produces the Fraunhofer diffraction pattern of the input information in two dimensions. This diffraction pattern has all the characteristics of a two-dimensional Fourier transform of the light distribution in the input information. Lens L_2 is called the Fourier transforming lens and F is referred to as the Fourier transform or spatial filtering plane. The diffracted light is recombined by imaging lens L_3 (which takes the inverse transform), to form the image of the input transparency in the output plane, W. Spatial filtering experiments are performed by inserting stops in the transform plane, F, to remove certain components from the frequency spectrum of the input transparency. The image in the output plane, W, will be altered due to this manipulation of the diffraction pattern.

The apparatus employed for optical filtering of a transparent object comprises several optical systems, including lenses. These lenses can cause disturbances in the interpretation of the resulting data. The various imperfections in these lenses, such as air bubbles, scratches, deposits of dust, give rise to parasitical diffraction effects, disturbing the reconstituted image. Further disturbances may be caused by imperfect assembly of the lenses and also by reflection of some rays of the light beam on their surfaces. These disturbances may be multiplied by the number of optical systems employed.

Figure 237 shows diagrammatically an apparatus which minimizes these disadvantages by employing a single, convergent, optical system.

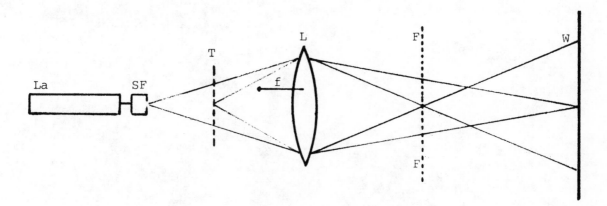

Fig. 237. Spatial filtering device employing a non-parallel laser beam and a single optical system.

The image on the screen, W, can be obtained by locating the input transparency, T, to be filtered in a divergent laser beam so that the light rays impinge on the transparency at relatively small angles of incidence. A convergent lens, L, is located between the transparency and the plane, F. The spectrum (diffraction pattern) of T is then formed in the transform plane, F. Filtering elements being interposed in the plane, F, the reconstituted filtered image forms on a screen W in the plane of the image of input transparency T conjugated with respect to the optical system, L. The source of the light (the pinhole in SF) and the transparency T should be at a distance from the optical system, L, greater than the focal length of L, which condition is necessary to obtain true images. The size of the spectrum in the plane F may be varied by simply displacing the input transparency T relative to the light source. It is thus possible to adapt the size of the spectrum (diffraction pattern) to the size of the filtering element, which is more advantageous than the reverse. As one filters the Fourier transform one can immediately see the effect on the reconstituted image which appears at W.

References:

A.B. Porter, Phil Mag v11 p154 (1906)

C.A. Taylor, and B.J. Thompson, J Sci Instrum v34 p439 (Nov '57)

L.J. Cutrona, "Optical and Electro-optical Information Processing,"
(MIT Press, Cambridge, Massachusetts, 1965), p83

D. Redman, Sci J p51 (Feb '68)

C.A. Pipan, Opt Spectra p50 (Nov/Dec '68)

R.A. Phillips, Am J Phys v39 p536 (May '69)

A. Fontanel, and G. Grau, British Pat. No. 1,168,698 (29 Oct '69)

H. Abbott, F.T. Johnson, R. Licata, and P. Oliver,
Am J Phys v39 p412 (Apr '71)

J.C. Brown, Am J Phys v39 p797 (Jly '71)

B.J. Pernick, Am J Phys v39 p959 (Aug '71)

G.R. Graham, Phys Educ p352 (Jly '72)

C.W. Curtis, and W.J. Van Sciver, Am J Phys v40 p1684 (Nov '72)

""""""""""""

Holography

Holography is a two-step process of optical imagery using a source of coherent light. First, a complex interference pattern is recorded on a photographic emulsion. This detailed permanent record of the necessary information is called a hologram. Second, the hologram is illuminated in such a way that part of the transmitted light yields a view of the original object recorded.

Since holography is an interference technique, the recording light must be coherent in space and time. Laser light has these qualities.

To make a hologram, the single beam of laser light is split into two components. One component is directed to a recording medium (usually a very fine grain photographic emulsion) and is called the reference wave. The other component is aimed at the object to be recorded. The second component is scattered or diffracted by the object and this scattered wave which is called the object wave is now allowed to fall on the recording medium. Since both the reference and object waves are mutually coherent, they will form a stable interference pattern when they meet in the plane of the recording medium. This interference pattern is a complex system of fringes. Different object waves (objects) will produce different interference patterns. The intensity and the relative phase of the light coming from the object is stored, upon exposure and development, on the hologram emulsion in a kind of optical code.

The second step in holography is the creation of an intelligible image from the hologram. This is called the reconstruction process. Holograms cannot be decoded at sight. The holograms must be illuminated with a laser beam which is similar to the original reference beam used to record the hologram. The "playback" laser beam is transmitted through the clean areas of the hologram resulting in a complex transmitted wave. The interference fringes recorded on the hologram act as a compex set of millions of microscopic diffraction gratings. Local variations in the contrast (amplitude) and spacing (phase) of the fringes produce local variations in the amplitude and direction of the diffracted playback beam. When the hologram is properly illuminated, the result is an undeviated wave and first-order diffracted waves on each side. One of the first-order waves exactly duplicates the original object wave. By viewing this reconstructed wavefront, one sees an image which appears as though the original object were in place and includes three-dimensional effects with parallax.

As mentioned before, a hologram consists of a complex distribution of microscopic clear and opaque areas corresponding to the recorded interference fringes. In addition, one can also observe clearly visible specks, whorls, spots and rings. These large-scale diffraction patterns are due to dust, dirt and scratches on one or more of the optical components used in making the hologram. These imperfections have negligible effect on the reconstruction.

In 1947 the theory of holography was conceived and developed by Dennis Gabor, a Hungarian physicist. For his discovery and pioneering work Dr. Gabor received the Nobel Prize in Physics in 1971. During the pre-laser days important contributions were made by G.L. Rogers, A. Lohmann and H.M.A. El-Sum among others. The invention of the laser in 1960 gave great impetus to holography research, especially when in 1963 E.N. Leith and J. Upatnieks invented the off-axis reference beam. This technique was made possible by the great coherence length of the helium-neon laser. In 1964, Leith and Upatnieks achieved an even more spectacular improvement by introducing diffused illumination of the object. With this technique the information from every object point was spread over just about the entire hologram, with the result that all of the object could be seen through any point of the hologram and seen also, as three-dimensional.

Since 1964 progress in holography has been very rapid. All over the world hundreds of researchers have been working to improve the techniques, as well as to find new applications. Among the present applications are optical-data processing, character recognition, holographic microscopy, particle-size analysis, 3-D displays, holographic interferometry, vibration analysis, non-destructive testing, flow visualization, contour generation, information storage, production of holographic lenses, gratings, filters and beam deflectors, and certain areas of diagnostic medicine.

Suggested Additional Readings:

E.N. Leith and J. Upatnieks, Sci Am v212 p24   (Jne '65)
A.E. Ennos, Contemp Phys v8 p153 (1967)
H.M.A. El-Sum, Sci & Technol p50 (Nov '67)
T.H. Jeong, Opt Spectra p59 (Nov/Dec '68)
M. Young, Am J Phys v37 p304 (Mch '69)
D. Gabor, Science v177 p299 (28 Jly '72)
J.D. Gaskill, Opt Engineering v13 pG15   (Jan/Feb '74) and
                               v13 pG119 (May/Jne '74)
T.H. Jeong, Am J Phys v43 p714 (Aug '75)
A.G. Porter and S. George, Am J Phys v43 p954 (Nov '75)
"Holography Using a Helium-Neon Laser," published by Metrologic Instruments, Inc., Bellmawr, New Jersey, 08030. This is a revised and expanded edition of Dr. T.H. Jeong's draft, "A Study Guide on Holography." The manual discusses the theory and physics of holograms and provides complete exposure and processing instructions for making holograms of many types. (1976)
E.M. Leith, Sci Am v235 p80 (Oct '76)

""""""""""""

Holograms have many fascinating properties. Inexpensive ready-made holograms on film can be purchased from several sources (for example, Edmund Scientific Co., and Jodon Engineering Associates, Inc.).

The following exercises require so-called 'transmission type' holograms. The hologram should be held at the edge to keep fingerprints from the hologram surface. The visible structures such as specks, concentric rings and large fingerprint-like shapes arise from dust particles and other scatterers in the optical system while recording the hologram. The pertinent information recorded on the hologram can only be seen under a microscope.

1. A two-dimensional real image can be projected on a screen by shining the undiverged laser beam straight through the back of the hologram. Fig. 238 illustrates this.

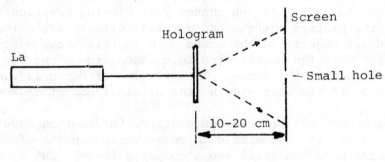

Fig. 238

Point the laser at a white card with a 2-mm hole in the center. Direct the laser beam through the hole. Turn off the lights in the room. Introduce the hologram into the beam path and observe that two images appear on the screen. The two images are two first-order diffraction patterns and are mirror images of each other. Move the hologram across as well as up and down in the laser beam and note changing perspective and parallax.

2. The most unique property of the hologram image is its three-dimensional appearance. Fig. 239 shows the setup. The laser beam is spread by placing a diverging lens (-5 to -10 mm f.l.) over the laser output window. The hologram is held in the laser beam 50-100 cm from the lens. At this distance the laser beam is wide enough to illuminate most of the hologram. Look through the hologram toward the laser at an angle as shown in Fig. 239. The hologram must be held in the correct position and viewed from the correct angle to see the image. Tip the hologram slowly as shown by the arrows and view it in various head positions. The image may appear upside down, distorted, or may not be visible at all. Turn the film around and view through the other side. With a little patience and experimentation you should be able to see a sharp 3-D image of the object or scene. Remember, the room must be darkened and you should not look at the hologram but through it.

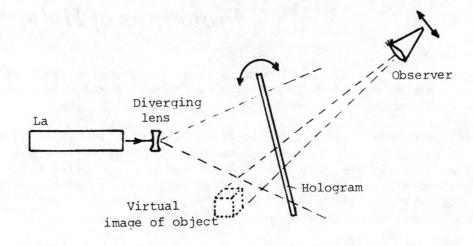

La

Diverging
lens

Observer

Virtual
image of object

Hologram

Fig. 239

3. While holding the hologram stationary in the correct position in
    reference to the laser beam, move your head slowly up and down and
    from right to left. As you change your viewing position the perspec-
    tive of the picture changes and parallax effects are evident between
    near and far objects in the scene. In addition, one must refocus
    one's eyes when the observation is changed from a near to a more
    distant object in the scene. The illusion of depth is the same as
    it would be if you were viewing the original object or scene.

4. Since each part of the hologram contains information about all parts
    of the object, it is possible to cover various parts of the hologram
    with a cardboard and still see a complete image. Cut a hole about
    one centimeter in diameter in the center of a cardboard and move the
    hole over the hologram while looking through it. Each portion of the
    hologram will reconstruct a slightly different view of the object.
    As the utilized hologram area is decreased the resolution of the image
    becomes poorer since resolution is a function of the aperture of the
    imaging system.   This effect can be shown by using holes with differ-
    ent diameters or an iris diaphragm.

5. A 180-degree rotation of a hologram around a vertical axis causes an
    inverted image.

6. A 180-degree rotation of a hologram about a horizontal axis also
    produces an inverted image.

7. A 180-degree rotation of a hologram about an axis perpendicular to
    the plane of the hologram does not cause an inversion of the image.

8. The wavefront reconstruction process does not produce negative images.
    If the hologram is copied by contact printing and then the copy viewed
    in the laser beam, the tonal values would still remain the same.
    Both the original hologram and its 'positive' print will reconstruct
    the same wavefronts just as the 'positive' copy of a diffraction
    grating will form the same spectrum as the original.

9. Microscopic examination of a hologram reveals that the emulsion used

in holography registers mainly two levels of density - opaque and more or less transparent. Still, the reconstructed wavefronts create an image with practically the same contrast rendition as the original object or scene and all the gray scale tonal properties of the original are preserved.

10. Point your camera at the hologram in the direction you were looking through the film. The camera focus will have to be altered for recording sharp foreground and background objects and the lens will have to be stopped down to increase the depth of field. The camera cannot tell the reconstructed waves from the original waves from the object. The camera will form an image of the original object - even though the object has long since been removed!

11. Another interesting property of holograms is that several images can be superposed ('stacked') on a single emulsion and each image can be recorded without being affected by the other images. In case of a double-image hologram the fringe pattern can be vertical for one exposure, and horizontal for another. Such ready-made double-image holograms are also available commercially. To view these images, the hologram is held in the diverging laser beam and adjusted until a bright and sharp image is seen in the space between the hologram and the laser. Once this is accomplished, the film is rotated $180°$ and the second image will appear.

12. The efficiency of reconstruction of a hologram image is not high. An amplitude grating is formed by the silver grains of the light sensitive emulsion. Such a thin absorption hologram has a maximum efficiency of about 6%. It is possible, however, to form holograms which have only phase variations. A thin phase hologram has a maximum theoretical efficiency of 33%. The amplitude grating is converted into a phase grating by various bleaching techniques described elsewhere in this book.

" " " " " " " " "

Requirements and Materials for Hologram Production

A)   The Light Source.

The light source used to record holograms must have a coherence length at least equal to the maximum difference between the object beam path length and the reference beam path length. Path length differences less than the coherence length are required to obtain holographic interference effects. Gas lasers offer by far the highest coherence length, and they are continuous in operation, which simplifies experimental procedure.

A laser intended for making holograms must operate in the TEM_{oo} mode. The lowest usable output is about 0.7-1.0 mW. To evaluate the coherence length of the laser to be used, a Michelson-type interferometer can be used as shown in Fig. 240.

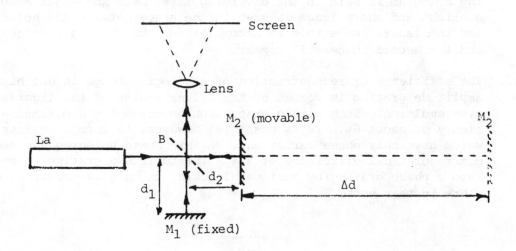

Fig. 240.   Determining coherence length.

The interferometer is set up for equal path lengths ($d_1 = d_2$). As the path difference Δd between the two mirrors is increased the fringe contrast will decrease. The minimum permissible fringe contrast will establish the usable coherence length which is equal twice the path length difference Δd.

For best results: 1) the total distance travelled by the reference and object beams from the beam splitter to the holographic film should be about the same, 2) the subject of the hologram must be small enough to fit within both the lateral and longitudinal coherence distances of the laser source, 3) the correct ratio of the two beam intensities must be established. The intensities of the interfering object and reference beams should be equal for maximum contrast in the interference fringe system. However, it is desirable to work on the straight line portion of the emulsion's D log E curve. To accomplish this, there must be no nulls in the interference pattern. For small objects the object beam should have 1/2 to 1/4 the intensity of the reference beam. A photometer (such as Metrologic 60-230) should be used to measure the beam ratio. The intensity of the object beam at the emulsion plane, after reflection from the object should be compared with the intensity of the reference beam at the emulsion plane. Room lights should be turned off and photometer readings taken by alternately blocking off one and then the other beam with a piece of black cardboard.

Another important factor in hologram production is the geometric arrangement of the object and reference beam as it affects spatial resolution. Most holographers prefer high speed emulsions which can resolve over 1,000 lines/mm. The angle θ between the two beams incident on the recording film should be between 30-45° with 60° as the maximum value.

B) Recording Media.

The detail that a hologram emulsion must record is very fine. This detail (interference fringes) is a function of the maximum angle (θ) between the object and reference beam. The fringe spacing in the hologram is $d = \lambda \sin \theta$. If we use a film whose resolving power is ρ lines/mm then the film cannot record those fringes whose spacing is less then $d_{min} = 1/\rho$, so that $\sin \theta_{max} = \lambda \rho$.

Kodak High Contrast Copy Film is one of the readily available conventional high resolution films. It has a resolving power in excess of 225 lines/mm. When using a He-Ne laser (λ=633nm) we find θ_{max} is approximately 8 degrees. Since we do not wish to view the reconstruction by looking almost directly into the laser beam, we prefer to use special photographic materials with very high resolution power: 1,000 to 2,500 lines/mm. If for example, θ=30°, the emulsion must resolve about 1,000 lines/mm. Since the recording emulsion must have a resolution better than the finest fringes to be recorded, one chooses the fastest film with adequate resolution power. Most of the holograms made since the early '60s have been recorded on Kodak Spectroscopic Plate Type 649-F which have a very slow-speed emulsion. Energy densities of 900 ergs/cm^2 were required to expose the Type 649-F emulsion with He-Ne lasers. This explains the requirement for high-power illumination or very long exposures. These plates are still available and useful in certain applications.

Kodak has recently introduced a high speed holographic emulsion particularly useful for general holographic procedures with low-power He-Ne lasers. The new products are identified as Kodak High Speed Holographic Plate, Type 131-02 and Kodak High Speed Holographic Film, SO-253. The SO-253 is ideally suited for experimental holography. This film is generally stocked in 25-sheet packages, 4x5", and 35mm x 150 ft. rolls, although other sizes are also available. These emulsions are available through Jodon Engineering Associates, Inc. and Newport Research Corp. in modest quantities, by mailorder.

The Type 131 Plates and the SO-253 Films are about 140 times faster than the previously mentioned Type 649-F Plates and they are 8 times faster than Agfa-Gevaert's 10E75 Plates when used with He-Ne lasers. The incident energy required to achieve optimum reconstruction brightness is about 5 ergs/cm^2 with He-Ne lasers. This exposure level can be reduced about 50% by adding 2 grams of Potassium Thiocyanate to each liter of developer. Resolution power is exceeding 1,500 lines/mm, corresponding to an angular beam separation (θ) of approximately 60 degrees.

This high speed emulsion is developed for 6 minutes in D-19 developer at 20C (68F). Following development, processing is continued with the following steps, all at 18.5 - 21C (65-70F)... Rinse: in Kodak Stop Bath SB-1a with agitation for 10-30 seconds. Fix: using Kodak Fixer or Fixing Bath F-5 with agitation for 5-10 minutes. Wash: with moderate agitation for 10 minutes. Rinse: in a solution of 3 pts. Methanol and 1 pt. water with agitation for 5 minutes. Wash: with moderate agitation for 5 minutes in slowly running water. Rinse: in Kodak Photo-Flo Solution for a few seconds to minimize drying marks and promote uniform drying. Dry: in a dust free atmosphere at room temperature. Note: The methanol rinse is required to remove a high level of residual sensitizing dye from the emulsion. The dye is distinctly blue in appearance and would greatly reduce reconstruction brightness when viewing the hologram with a red-emitting laser. Some investigators have reported relatively rapid bleaching of this dye under exposure to fluorescent lights or daylight!

Over the range of from 1 to 10^{-4} secs, this emulsion shows virtually no reciprocity effects. For exposure durations of 10 to 100 seconds, exposures should be approximately 25% greater than those calculated on the basis of 1 second or shorter exposure time. The emulsion exhibits significant latent image fading during the hours following exposure. For example, an exposure sufficient to yield a density of 1.0 when the film is processed immediately, will result in a density of 0.8 if processing is deferred for one hour. Unexposed films and plates should be stored in a cool place (21C = 70F or lower) in the original sealed package. Total darkness is recommended when handling these emulsions. Greater care is required in shielding the emulsions from laser radiation extraneous to the actual holographic exposure than one may be accustomed to providing with materials previously available.

Bleaching of amplitude holograms is generally used for forming high diffraction efficiency phase holograms. The most common procedure is to convert the silver grains formed after developing into a transparent compound with refractive index different from that of the gelatin. Thus, in a phase hologram, the recording takes the form of spatial variations in optical path length instead of spatial variations in optical density, as in the original amplitude hologram. It is interesting to note that good bleached holograms can be made from 2x to 10x overexposed amplitude holograms. A phase modulated hologram may be as high as 10-times more efficient than an amplitude hologram.

Three simple methods for bleaching developed and fixed holograms is given below:

Bleach Formula No. 1

    Water...   1 liter
    Potassium Dichromate...   9.5 grams
    Concentrated Sulfuric Acid*...   12 milliliters

*Warning: Always add the sulfuric acid to the solution, slowly, stirring
        constantly!

Bleach for 2-4 minutes; rinse in running water for 5 minutes; immerse in Kodak Photo-Flo (2 ml per liter) for about a minute and dry at room temperature.

Bleach Formula No. 2

    Potassium Ferricyanide...   45 grams
    Potassium iodide...   25 grams
    Water...   1 liter

    Use full strength. Bleach until emulsion has cleared. Rinse one minute
in water. Rinse one minute in isopropyl alcohol. Air-dry at room temperature.

Graube's Bleaching Method

    The processed and dried hologram is placed in a container with a few drops
of bromine in the bottom. The container is ealed airtight. In a few minutes
the emulsion is bleached with excellent diffraction and keeping qualities.
The emulsion remains dry during the process. Bromine is volatile and toxic.
Handle with care: bromine can burn the skin. Provide good ventilation.

Suggested Additional Readings Regarding Phase Holograms and Bleaching Methods:

W.T. Cathey, Jr., J Opt Soc Am v55 p457 (1965)
J.H. Altman, Appl Opt v5 p1689 (1966)
V. Russo and S. Sottini, Appl Opt v7 p202 (1968)
H.M. Smith, J Opt Soc Am v58 p533 (1968)
J. Upatnieks and C.D. Leonard, Appl Opt v8 p85 (1969)
J. Upatnieks and C.D. Leonard, J Opt Soc Am v60 p297 (1970)
D.H. McMahon and W.T. Maloney, Appl Opt v9 p1363 (1970)
R.L. Lamberts and C.N. Kurtz, Appl Opt v10 p1342 (1971)
A. Graube, Appl Opt v13 p2942 (1974)
R.L. Lamberts, Appl Opt v11 p33 (1972)
N.J. Phillips, and D. Porter, J Phys E: Sci Instrum v9 p631 (Aug '76)

Films for Holographic Reconstruction Photography

    The films listed below are well suited to holographic reconstruction
photography using He-Ne lasers:

Kodak Tri-X Pan: General purpose, fast film; ASA 400; 35-mm magazines.

Kodak Type 2475 Recording Film: For low level illumination or short exposure
    times; ASA 1000+; 35-mm magazines.

Kodak Type 2485 Recording Film: Extremely high speed; ASA 800-8,000 with
    special developer; 35-mm x 150 ft. roll.

Kodak Type 2474 Linagraph Shellburst: Good exposure latitude; moderate speed
    -ASA 200; 35-mm x 150 ft. roll.

Kodak Type 4155 Contrast Process Pan Film: A high-resolving power, fine-grain,
    high-contrast sheet film; ASA 100; 4x5", 25 sheets per box.

Kodak Type SO-410: Extremely fine-grain, extremely high-resolving power,
    moderate speed; ASA 80-160 contrast dependent; 35-mm magazines.

    Full technical specifications and processing instructions regarding these films can be obtained from Jodon Engineering Associates, Inc. (Jodon No. KLP-R6), or Eastman Kodak Company.

C) System Stability

Vibration is a practical difficulty encountered in making holograms. Since a hologram is an interference pattern, nothing in the entire holographic apparatus must be permitted to move more than a quarter wave-length during the exposure. Vibrations of the work surface, air currents, acoustic waves and temperature fluctuations can obliterate the interference patterns. The vibration problem becomes more and more severe as the angle (θ) between the object and reference beam is increased. A long-legged Michelson interferometer set up on the work surface can be used to monitor surface stability. Figure 241 shows the arrangement.

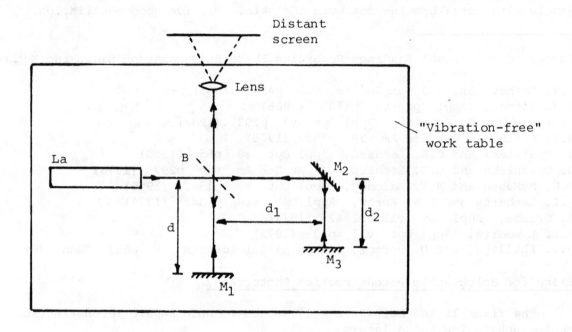

Fig. 241. Determining system stability.

The component parts are spread out to cover as much area as possible on the work table. By making $d = d_1 + d_2$ maximum fringe contrast will be obtained. The fringes can be observed on a distant wall or screen. Vibrations of the table and the optical components will cause the fringes to shift. A shift of one-half interference fringe width across the recording film during exposure will prevent the hologram from yielding a reconstruction. A movement of only one-eight of a wavelength in a direction normal to the film will impair reconstruction. Thus, in addition to a vibration-free work surface very rigid component mountings must be used.

It is most important to minimize vibrations that alter the distance between the object and the mirror which re-directs the reference beam to the film. Vibrations between the recording film and the laser, and between the laser and the object have much less detrimental effect on the quality of the hologram, but they too must be minimized.

Construction of a vibration isolation surface for holography is quite simple and need not be expensive. The entire assembly should preferably rest on a cement floor in a garage or basement. If the holography equipment is to be set up on a laboratory bench top or on a heavy work table, the legs of the table should sit in buckets of sand.

First, four rubber balloons with air valves are placed on the work table, arranged so that they will be under the corners of the optical table carrying the various optical components. The balloons should be positioned so that the inflating valve and most of the neck is exposed. Next, a piece of square hardboard (plywood or masonite) is placed over the balloons and a thick foam rubber square on top of the hardboard. The optical table (a heavy steel plate) is thus supported at each corner by an "isolation sandwich." (see Fig. 242, side and top views).

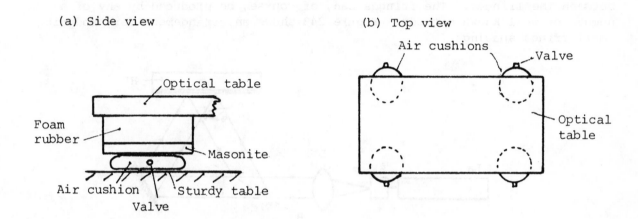

(a) Side view

(b) Top view

Fig. 242. Vibration isolation.

Finally, the air cushions (balloons) are inflated very slightly, so that the optical table just "floats" without touching the work table at any point. An air cushion thickness of approximately one centimeter (3/8") will be sufficient. Optical table, foam pads, rubber balloons with air valves and Masonite sheets are available from Metrologic Instruments, Inc. The triangular optics benches carrying the various components are easily attached to the optical table by cementing several strips of magnetic tape across the width of the benches and holding them magnetically in position.

References:

G.L. Rogers, J Sci Instrum v43 p677 (1966)
J.P. Nicholson,, Phys Educ p41 (Jan '76)
J.R. Duckett, G.F. Lothian and J.F. Parsons, Phys Educ p294 (Jne '76)

" " " " " " " " " "

Interferometric Method of Producing Optical Diffraction Gratings

Interferograms as well as holograms are recordings of the interference pattern created by an object wave and a reference wave.

Optical diffraction gratings can be produced by a holographic method in which the photosensitive layer is exposed to a set of interference fringes produced by two coherent beams of light intersecting at an angle. By varying the angle between the two beams it is a simple matter to modify the spacing between the fringes. The fringes can, of course, be produced by any of a number of well known methods. Figure 243 shows an arrangement for producing small fringe spacings.

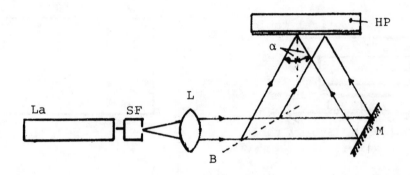

Fig. 243. Method of producing closely spaced fringes.

The beam from a He-Ne laser is filtered (SF), expanded and then collimated with a lens (L). This beam is divided into two collimated beams by means of a beam splitter (B) and are recombined on a photographic plate (HP) with the help of a front surface mirror (M). The fringe spacing depends upon the wavelength of the generating laser light and upon the angle (2α) between the beams. This angle can easily be modified by varying the position of the beam splitter and the mirror. The separation between successive fringes, that corresponds to the grating constant σ, is given as $\sigma = \lambda/(2\sin\alpha)$. If, for example, $\alpha = 8°$, the grating constant is about 1/500mm for the He-Ne laser. Kodak holographic films and plates are capable of recording such linear interference fringes since their resolution capability exceeds 2,000 lines/mm.

Figure 244 depicts a Wollaston prism arrangement suitable for large fringe spacings.

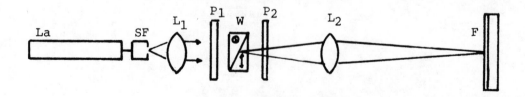

Fig. 244. Method of forming broadly spaced fringes.

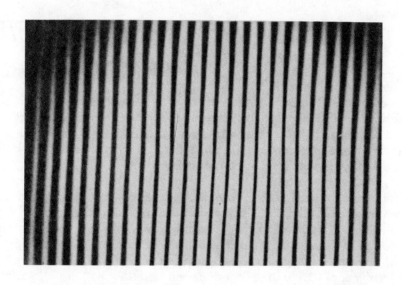

Fig. 245. Straight line fringes (enlarged).

The Wollaston prism placed between polarizers P_1 and P_2 produces sinusoidal fringes which are localized in the interior of the Wollaston prism. When the polarizers are changed from a parallel to a crossed position or vice versa, there is an effective displacement of fringes by half a period. L_1 is the collimator lens and L_2 images the fringes on the photosensitive layer F.

References:

M.H. Horman, Appl Opt v4 p333 (1965)
O. Bryngdahl, and A.W. Lohmann, J Opt Soc Am v58 p141 (Jan '68)
C.P. Grover, S. Mallick, and M.L. Roblin, Opt Commun v3 p181 (May '71)
J. Heidenhain, H. Burhardt, and H. Kraus, British Pat.No. 1,261,213 (26 Jan'72)
T.T. Stapleton, U.S. Pat. No. 3,695,749 (3 Oct '72)
M. Yoshida, K. Yoshihara, and K. Kamiya, Science of Light, v22 p146 (1973)
M.C. Hutley, J Phys E: Sci Instrum v9 p513 (Jly '76)

"""""""""""

Optical Method for Producing Zone Plates

The classical Fresnel zone plate, with its alternate opaque and transparent zones, acts as a lens with diffraction rather than refraction serving to deviate the light rays. The Fresnel zone plate is an unusual lens, however, in that it possesses multiple foci and also because it behaves simultaneously as a diverging and a converging lens.

The classical zone plate with alternate opaque and transparent zones has an amplitude transmission which is square-wave in form. A zone plate in which the boundaries between light and dark regions are gradual rather than abrupt can be produced by photographic (holographic) recording. The transmission function of a photographically produced zone plate approaches a sine-wave form. This kind of zone plate has been named a 'Gabor Zone Plate.' It possesses a single positive and a single negative focus.

The presence of single real and virtual images in holographic reconstructions suggests a strong resemblance between the holographic process and the imagery of a sine-wave (Gabor) zone plate. This similarity led G.L. Rogers in 1950 to suggest that a hologram might be regarded as a generalized zone plate. In 3-D holographic practice, each point on the object interferes with the reference beam to form an interference pattern resembling a Fresnel zone plate of circular fringes. The hologram is then a collection of individual, or part of individual, sine-wave zone plates. The zone plates have various principal focal lengths, since each scattering point in the object is, in general, a different distance away from the hologram plate than all the other points. Each of the constituent zone plates reconstructs according to its own focal length, thus reconstructing 3-D real and virtual images.

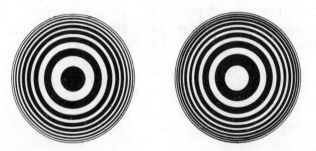

Fig. 246. Fresnel zone plates with black & white centers.

A point hologram is the fringe pattern formed on a photographic plate when a spherical wavefront interferes with a plane wavefront, or with another spherical wavefront of different curvature. The photographic recording of a point hologram is a simple method of making large Gabor zone plates.

Sine-wave zone plates can be produced by photographing a Newton's rings interference pattern. The experimental setup can be seen on page 126 (Fig. 150). Kodak Panatomic X, Kodak Photomicrography Monochrome Film, Type SO-410, or a high-resolution holographic emulsion (for example, SO-253) can be used. Figure 247 shows enlarged prints from such a zone plate photographed on 35-mm film. The rings were photographed: (a) camera centered on axis, and (b) camera turned off center, exposing more rings to view. Point holograms of any desired focal length can be made by interfering two wavefronts of different curvature.

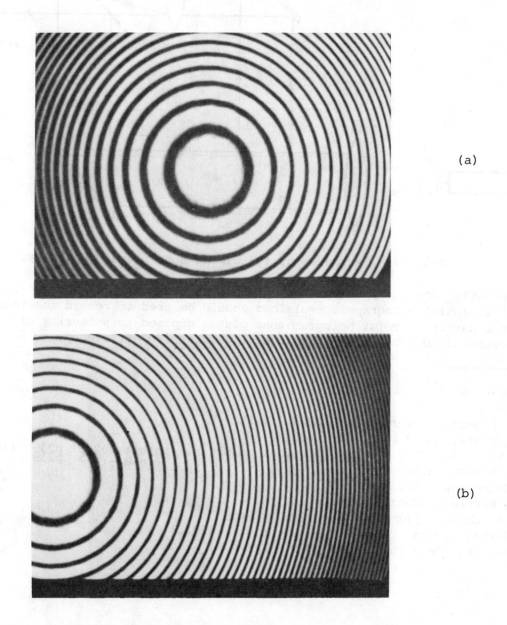

(a)

(b)

Fig. 247. Newton's rings interference patterns.

Point holograms of various focal lengths can be made using a Mach-Zehnder interferometer by interfering a spherical wavefront with a planar reference wave. The experimental arrangement is shown schematically in Fig. 248.

Positive lens, L_1, collimates the divergent spatially filtered laser beam. Another positive lens, L_2, is inserted in the object field.
Thus, the zone plate itself is a fringe pattern created by the interference

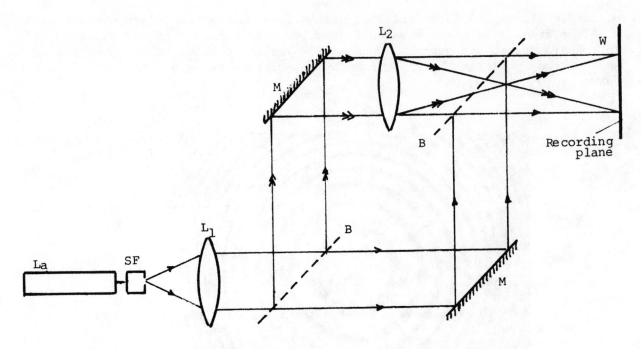

Fig. 248. Arrangement for making zone plates optically.

of mutually coherent spherical and plane wavefronts from the He-Ne laser. High resolution holographic emulsions should be used to record the interference pattern. Point hologram zone plates exposed to a density of approximately 0.4-0.5 yield the highest diffraction efficiency.

References:

G.L. Rogers, Nature v166 p237 (1950)
O.E. Myers, Am J Phys v19 p359 (1951)
M.J.R. Schwar, T.P. Pandya, and F.J. Weinberg, Nature v215 p239 (1967)
M.H. Horman, and H.H.M. Chau, Appl Opt v6 p317 (1967)
E. Champagne, Appl Opt v7 p381 (1968)
W.G. Ferrier, Contemp Phys v10 p413 (1969)
H.H.M. Chau, Appl Opt v8 p1209 (1969)
M. Parker Givens, Am J Phys v40 p1311 (1972)
J. Higbie, Am J Phys v44 p929 (Oct '76)

"""""""""""

Low-budget Holography

Contrary to general belief, holograms are not difficult to make and the equipment need not be expensive.

Holography experiments are in the category of optical interferometry and, as such, extreme mechanical stability of all optical components is a basic requirement for success. Most of the failures in hologram production can be traced to vibrations which produce relative movement between the components of the optical system. The required optical stability for hologram exposure must be provided by an isolation system in which the optical table is 'floated.' An inexpensive, relatively large, vibration-free working surface (optical table) is described elsewhere in this book.

A low-power (0.5 to 2.0 mW) continuous wave He-Ne gas laser operating in the TEM_{oo} mode is used to take holograms of small, diffusely reflecting objects. These lasers, thanks to quantity production, are not expensive.

To keep exposures short while using low-power lasers one must use the fastest holographic emulsions available and suitable for a particular project. In the early '60s, and for years thereafter, the pioneers of holography used Kodak spectroscopic plates (Type 649-F) as the recording medium. These plates are still used today. Their resolving power is very high, - more than 2,000 lines/mm, - but the emulsion is of very low sensitivity. In the '70s Kodak introduced a high-speed holographic film, Type SO-253. It is available in both 35-mm and 70-mm widths on 150 foot long rolls and also in sheet film form. This high-speed holographic emulsion is capable of resolving 1,250 lines/mm. As a comparison, if a holographic setup requires 2 to 3 minutes of exposure on Type 649-F, the SO-253 film will record a good hologram with a one-second exposure. High emulsion speed allows short exposure times and that in turn minimizes the problem of vibration. The SO-253 is recommended for the holography experiments described on the following pages.

In certain instances acceptable holograms can be produced on Kodak High Contrast Copy Film, Type 5069. It is a standard, easily obtainable film that is supplied in 35-mm rolls of 36 exposures. Its speed is about the same as that of SO-253 (for He-Ne light) but its maximum resolving power is 630 lines per millimeter.

The smallest fringe spacing in a hologram determines the maximum emulsion resolution required. The fringe spacing is described by the expression, $d = \lambda/\sin\theta$, where θ is the angle between the reference and object beam at the film. θ is often referred to as the 'design angle.' Once the fringe spacing is calculated, a holographic film of higher resolution should be chosen to minimize grain noise. It is common to characterize emulsions by a factor "p" which is the number of lines per millimeter the film can record. For a He-Ne laser ($\lambda=633nm$), $p = \sin\theta/\lambda = 1580 \sin\theta$ mm^{-1}. Kodak High Contrast Copy Film can be used with good results for design angles (θ) up to 25 degrees. For larger design angles the SO-253 film must be used.

A) Two-dimensional holography.

A two-beam interferometric process is used in producing the hologram, as shown in Fig. 249.

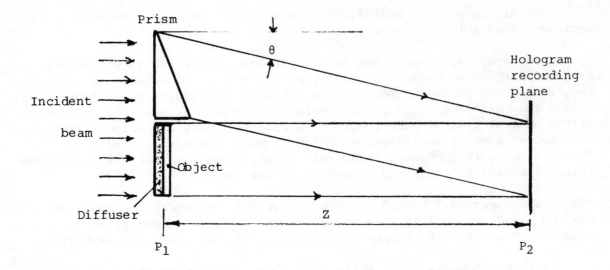

Fig. 249. Prism technique for producing a two-beam hologram.

The object (transparency), located at plane P_1 is illuminated with collimated laser light. A Fresnel diffraction pattern of the object transparency is formed at plane P_2. Adjacent to the object is a prism (a Herschel wedge of approximately 15-75-90 degrees, or a 30-60-90 degree prism). The portion of the incident beam which impinges on the prism is deviated through an angle θ and thereby the two beams are brought together at a distance z to produce a Young's fringe pattern. A photographic (holographic) emulsion at P_2 records the resultant pattern, thereby producing a hologram from which a reconstruction of good quality can be obtained.

In the absence of the object transparency, a uniform fringe pattern is produced by the interference between the two beams intersecting at the film plane. When the transparency is present, its Fresnel diffraction pattern modulates this fringe pattern. The amplitude portion of the Fresnel diffraction pattern amplitude-modulates the fringes and the phase portion produces a phase-modulation (spacing modulation) of the fringes.

The fine-line structure of the hologram causes it to act like a diffraction grating. When illuminated with monochromatic, spatially coherent light, the hologram produces a zero-order spectrum and a pair of first-order spectra. One of the first-order spectra forms a real image, and the other a virtual image.

Good reconstructions are possible over a wide range of fringe contrasts. The contrast of the fringe pattern can be controlled by the use of an attenuating filter placed in one of the beams. The reconstructed image has the same contrast as the original object, regardless of the gamma of the photographic emulsion. The reconstructed image is a positive. If the original hologram is copied so as to produce a negative then this negative hologram also produces a positive reconstruction.

As shown in Fig. 249, a diffusing element such as an opal glass, ground glass, or anti-glare glass is placed between the incident beam and the object transparency, thus causing the object transparency to be illuminated with diffused light.   The diffused light impinging on the transparency has a phase and amplitude which vary randomly from point to point, but these amplitude and phase relations are time invariant.   Because of the diffuser, each point on the transparency illuminates the entire hologram emulsion.   Thus, the observer sees the reconstructed image as though it were illuminated by a diffuse source (since the random wavefront originating at the diffusing plate is also reconstructed).   In the diffused-illumination hologram local imperfections in the optical elements (scratches, dust, etc.) are removed from the reconstructed image.   Such holograms can be abused (scratches, dirt, fingerprints, etc.) without noticeable deterioration of image quality.

The prism method of producing interference fringes is not the optimum was because the prism will introduce astigmatism, unless the incident beam is accurately collimated.   Mirrors are preferred for deviating the refernce beam.   Such an arrangement is shown in Fig. 250.

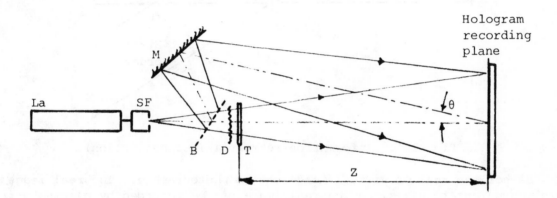

Fig. 250.   Mirror technique for producing a two-beam hologram.

The laser beam is filtered and made divergent with the spatial filter (SF).   The diverging beam is directed to a partially silvered (aluminized) mirror (B) which performs a beam splitting operation.   A totally reflecting front-surface mirror (M) is placed in the path of the reflected (reference) beam to direct the beam to the hologram emulsion (H).   The diffused light transilluminates the object transparency (T).   The light passing through the transparency produces a beam of scattered light which carries the Fresnel diffraction pattern of each point on the transparency, some of which is captured by the hologram film.   High-resolution holographic film is used to record the complex combination of multiple Fresnel patterns and interference fringes.

After the photographic film is developed, reconstruction is accomplished according to the diagram of Fig. 251.   The hologram H is illuminated by an incident beam of coherent light and a real image forms at a distance Z on one side of the hologram, and a virtual image forms at a distance Z on the other side of the hologram.   The fine line structure of the hologram causes the hologram to act like a diffraction grating, producing a first-order pair of diffracted waves, as shown in Fig. 251.   One of these produces the real image, occurring in the same plane as a conventional real image, but displaced to an off-axis position with respect to the incident illuminating beam through the angle θ.   The angle θ and distance Z will be the same in the reconstruction process as they were in the hologram-forming process if the same wavelength of light is used in both instances.

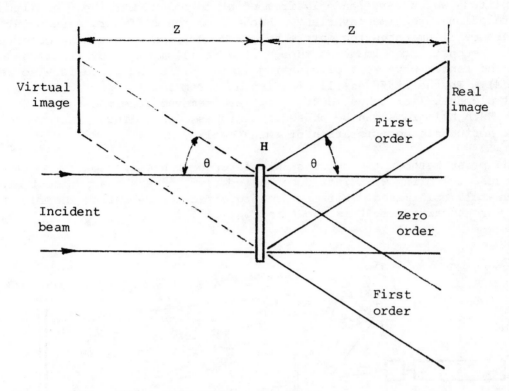

Fig. 251. Information retrieval (reconstruction).

Either the real or virtual image can be photographed. The real image is more convenient to use since the real image can be recorded by placing a film at the image position, determined by the distance Z and the angle θ, thus avoiding the need for a lens.

B) Three-dimensional holography.

The two-beam method is also effective with three-dimensional opaque objects. Holograms made from diffusely reflecting objects have all the properties of the diffused-illumination 2-D hologram previously described and, in addition, 3-D objects reconstruct as 3-D images.

The basic process used to photograph a solid three-dimensional object by wavefront reconstruction method is indicated in Fig. 252. The expanded laser beam illuminates the object. The diffusely reflecting object reflects light and exposes the ultrahigh-resolution holographic film (H). A portion of the incident beam is reflected to the film by a front-surface mirror located adjacent to the object. This provides the reference beam against which the phase of the object beam is compared. The incident beam is reflected by the mirror through an angle θ. The interference of the scattered object beam and the reference beam produces a hologram on the holographic film. After the film is developed, the semi-transparent hologram is placed in a beam of coherent light, as shown in Fig. 253 to view the reconstruction.

The hologram may be replaced in its original (recording) position and

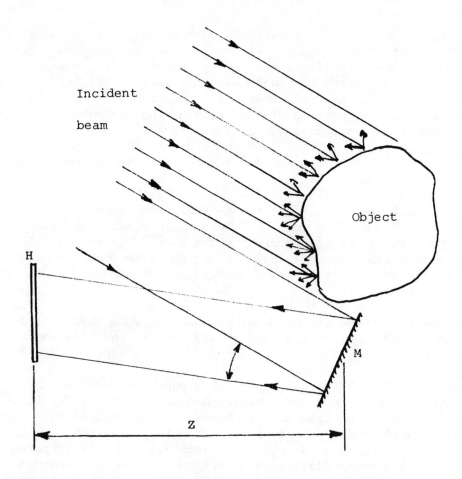

Fig. 252. The principle of 3-D holography (recording).

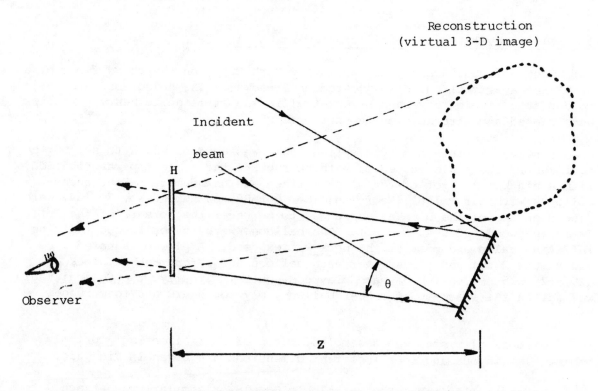

Fig. 253. Principle of 3-D holography (reconstruction).

the object removed. A viewer looking through the hologram as if it were a window (not into the laser beam!) will see an undistorted view of the object, just as though it were still present. This is the virtual image. In addition, a real image will be formed on the observer's side of the hologram but the image will appear fuzzy and distorted. A distortion-free real image can be formed by changing the position of the illuminating laser beam so that it appears to come from a mirror image of the reference beam in relation to the hologram. The undistorted, real, 3-D image of the object then forms in front of the hologram and appears suspended in space between the observer and the hologram.

The virtual image has all the three-dimensional appearance of the object or objects. If the observer moves his head while viewing the reconstruction, there will be a change in perspective of the image. There will also be a parallax effect between the near and far objects. The observer will have to refocus his eyes when shifting observation from near to far objects in the reconstructed scene.

While the real image can be observed, most viewers have difficulty in coordinating the eyes when doing so. In addition to the reversed parallax of pseudoscopic imagery, the real image suffers from depth inversion.

To sum up, it is the virtual image that can be easily seen by an observer. If the virtual reconstruction is photographed with a camera, it is necessary to focus the image on the film and stop down the camera lens to achieve a reasonable depth of field. The real image focuses to an image on the other (front) side of the hologram an equal distance (Z) from the hologram. This real image can be used to illuminate a screen. It can be recorded by placing a photographic film at some suitably chosen position, a compromise position, since the image is three-dimensional and the film is two-dimensional.

3-D Holography: Experimental Arrangement No. 1

Good holograms were produced in New York City, on the third floor of a high-rise apartment building located on Broadway. The holograms were recorded at a time of day when street traffic was still quite heavy. Windows were closed and airconditioning shut off.

A heavy, living-room type coffee table was used on which to set up the equipment. The table rested on a thick rug and the table top was covered with a blanket to protect the glass surface against damage. Four rubber balloons with air valves (Metrologic No. 60-608, or Edmund No. 71-184) were placed on top of the blanket, arranged to be under the corners of the Optical Table (plywood: 90 x 40 x 2 cm). The balloons were so positioned that the inflating valve and most of the neck was exposed. A plywood square (20 x 20 x 2 cm) was placed over each balloon and a foam rubber block (20 x 20 x 5 cm) on top of each playwood square. Finally, the Optical Table was set on the stack-up of rubber balloon, plywood board and foam rubber blocks.

The optical system was assembled on top of this 'floating' Optical Table. The layout and some important dimensions are given in Fig. 254.

The spatial filter (SF = Metrologic 60-618 or similar) was attached to the laser (Metrologic ML-968; 0.9 mW), and adjusted to obtain a 'clean'

diverging beam. To further expand the already diverging laser beam, a convex
mirror was placed in the beam path. The convex mirror (Metrologic 60-681) is
the same one used in the "360-degree Holography" experiment. Instead of this
mirror, a small plano-convex lens of 35-40 mm focal length can be used with
the convex surface aluminized.

The small object carrying table (target table) and the front-surface
reference mirror (about 5x7 cm) were affixed to optics carrying mounts and
then firmly mounted on one short (25-30 cm) triangular optical bench. The
beam expanding convex mirror and the film-holder (home made, or a 35-mm SLR
camera without the lens) were carried on another similar optical bench. The
whole optical system was very sturdy and compact.

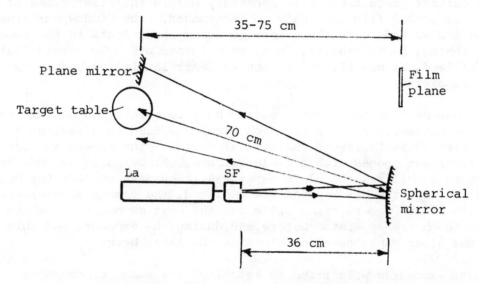

Fig. 254. Simple arrangement for making 3-D holograms.

Once everything was arranged on the Optical Table, the air cushions
(rubber balloons) were inflated very slightly with a bicycle pump. A balloon
thickness of approximately one centimeter (3/8") was found to be sufficient.
Too much air in the balloons had a deleterious effect on the vibration isolation capability of the system.

Next, the 'target' to be holographed was placed on the object carrying
table. Miniature red and white chessman, dice, an N-gauge model train car,
ping-pong ball, tiny figurines, the ball-like element of an IBM Selectric
typewriter and similar items were used as targets. The triangular optical
bench, the sliding object table mount, and the sliding reference mirror mount
were then adjusted so that the expanded laser beam fully illuminated the
target and partially the mirror.

The light reflected to the film by the reference mirror should be about
four times brighter than the light scattered to the film by the target. In
practice, an illumination ratio from 3:1 to 10:1 will produce fine holograms
and we have produced acceptable holograms with 100:1 ratio. The laser was
covered with black cloth to make sure that the only light emanating from the
laser came from the pinhole of the spatial filter. A Metrologic 60-230 laser
power meter was used to check the beam illumination levels at the film plane.
With the room lights off, the light from the mirror and from the target were

measured by alternately inserting a piece of black paper to block one of the light beams. The beam balance can be changed by moving the object (target) one way or the other, or by moving the mirror either toward the edge of the expanded laser beam or further away from the laser if the reference beam is found to be too intense. Note, however, that the length of the beam paths from the mirror and the target must remain equal within a few centimeters. This is much more important than to achieve optimum illumination ratio. The arrangement shown in Fig. 254 always produced good holograms. The angle of the plane mirror must be carefully adjusted to make sure that the reference beam covers the film. When the camera (film) was 35 cm away from the target and reference mirror, a 'close-up' hologram resulted. When the distance was set at 70 cm, a 'wide-angle' image was obtained.

Some further precautions were necessary before the first frame of the 35-mm roll of SO-253 film was ready to be exposed. The window curtains were drawn and we made sure there were no other light leaks in the room. Air-conditioning, heating units, fans, were turned off. To assure that the laser stabilized, it was allowed to run at least 15 minutes before making the first exposure.

There was no smoking, no unnecessary body movement, and no talking while the film was exposed. A strip of black paper was used as a shutter to block the laser beam where it leaves the spatial filter. The camera shutter (set on "T" = time) was opened with a cable release. Allowing 10 seconds for everything to stabilize, the black paper was removed to unblock the beam. After the exposure the black paper was replaced, the camera shutter closed, and the film in the camera transported for the next exposure. Nothing was allowed to touch the apparatus before and during the exposure and this included the black paper used for blocking the laser beam.

Several exposures were tried to establish the range of exposures which would produce acceptable holograms. Using a 0.9 mW laser (Metrologic ML-968), a spatial filter (Metrologic 60-618) and Kodak SO-253 film in the configuration shown in Fig. 254, we obtained good results with 1/4, 1/2 and 1 second exposures. These values are 'guesstimates' since the exposures are made by removing the black paper from the laser beam and then quickly replacing it to block the beam again. The exposed film was developed in D-19 and Kodak instructions were followed, including the methanol bath. After drying, the hologram was placed in the position the film occupied during the exposure. The objects used as target were removed from the target table. By looking through the hologram toward the empty target table the reconstructed 3-D image of the target appeared. If no reconstruction occurs, the first thing to do is to turn the film upside down and look again.

Troubleshooting: If no image can be reconstructed from the developed film, the most probable reason for failure is that something moved during the exposure. Make sure that all parts on the Optical Table are mounted very firmly. If the target carrying table appears in the reconstruction but not the target itself, it means the target moved during the exposure. Cement the target to the table and allow 24 hours to pass for the cement to fully cure. Also check the following: 1) Is the beam balance ratio correct? 2) Is the beam path difference within 5-10 cm? 3) Does the reference beam cover the film? 4) Is all stray light from outside and from the laser eliminated? 5) Is the exposure time correct? (When dry, good holograms appear as light or medium gray transparencies when examined in white light.) 6) Was the film properly processed? (Proper sequence, time- and temperature-control and cleanliness.)   7) Was processing delayed after exposure? (When processing is delayed after exposure, "latent-image fading" occurs and this is quite noticeable with microfine-grain holographic emulsions. Most density loss occurs during the first few hours after exposure.) 8) If you find the reconstructed image too far to one side of the hologram, adjust the angle at which the reference beam strikes the film, and the angle between reference and target beam.   9) If a 35-mm camera is used as the "film-holder", make sure the camera opening does not vignette the reference beam.   10) Over-exposed holograms can be saved by bleaching the silver image (Phase holograms require a somewhat higher exposure level).

3-D Holography: Experimental Arrangement No. 2

    This system will allow the serious holographer to produce superior quality, larger sized (4 x 5") holograms and to perform advanced experiments in stress- and vibration-analysis. The experimental configuration is shown in Fig. 255, the layout in Fig. 256. The relative placement of the components shown should be adhered to while making the first few holograms. Note that both the refer-ence beam path and the target beam path are 35 cm, measured from the beam splitter to the center of the film (plate). The optical system is assembled on the "floating" Optical Table as previously described. Since this is a different layout and a heavier setup, the size of the optical table is about 60 x 80 cm and is made of steel instead of playwood (for example: Metrologic No. 00152; part of the 60-635 Advanced Holography System).

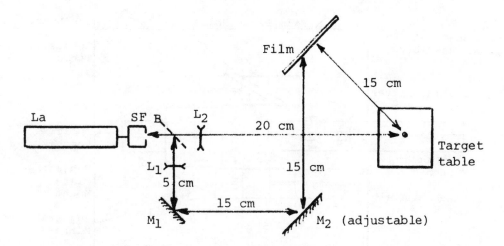

Fig. 255. Experimental arrangement #2 for 3-D holography (not to scale).

    Seven pieces of equipment are needed for this setup: (not counting the optical benches and component carrying mounts) a laser, a spatial filter, a beam splitter assembly, two reflecting mirrors, a film or plate holder, and a platform for the target.

    The laser should have an output of at least 2 mW and operate in the TEM_{OO} mode (Metrologic ML-650 or the modulatable model ML-969). The beam splitter assembly includes a plate glass beam splitter and a pair of diverg-ing lenses. The beam splitter is a small piece of highly polished optical glass having exactly parallel surfaces and a thickness of 5-6 mm minimum to separate the reflections from the front and rear surfaces. A Herschel wedge (Edmund No. 30-265) is a satisfactory beam splitter and due to the wedge angle it is quite easy to separate (block) one of the reflected beams. One may use a cube beam splitter (Edmund No. 30,329) which is somewhat more expensive. Quite a bit more expensive is the variable density beam splitter (Edmund No. 41,960). This is a variable density film of Inconel coated on a 25x75 mm glass plate. The transmitted light can be varied from 1% to 96%, allowing one to conveniently adjust the reference and target beam ratio. The diverging lenses are about 10 mm diameter and 8-9 mm focal length (such as Edmund No. 94,755).

The two reflecting mirrors are front-surface optically flat mirrors. One is a fixed mirror, the other one is mounted adjustably. The size is approx. 102x127 mm (4" x 5", Edmund No. 40,042).

The film or plate holder can be home-made, or a commercial unit (Metrologic No. 60-616). The target platform is a horizontal plate made from metal or wood, 15 cm square, attached to a vertically adjustable pin mount.

The laser should be positioned so that the spatial filter is as close to the beam splitter as possible without danger of touching it. The relative placement of the components is indicated in the illustrations.

The importance of stray light protection, the balance of beam intensities, and the equalization of beam path lengths were already dealt with in the previous section (Experimental Arrangement No. 1).

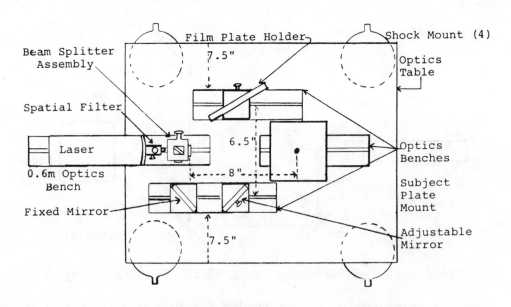

Fig. 256. Component layout for arrangement #2.

A convenient way to shutter the laser is to plug its power cord into an outlet box that is controlled by a switch. In this manner, the laser may be turned on and off without making physical contact with it. It is important, however, to have the laser on during the preparatory work for at least one half-hour before the first exposure is made to allow it to stabilize. Exposure time must be determined experimentally, since it depends upon the type of film or plate being used; color, size and reflectivity of the target;

distance from target to film; and relative beam intensities. Kodak High Speed Holographic Film SO-253 and the Kodak High Speed Holographic Plate 131 (both 4" x 5" size) are not inexpensive. Since Kodak's High Contrast Copy Film, Type 5069 has about the same sensitivity for He-Ne laser light as the above-mentioned emulsions, it is helpful to cut a piece of this 35-mm HC Copy Film, tape it to the film/plate holder and make trial exposures. While no hologram recording is expected on this film in this configuration (the design angle θ it too high), the density of the developed HC Copy film will be a helpful guide to those who previously experimented with 2-D holography but skipped Experimental Arrangement No. 1. Those readers who obtained good holograms with the Experimental Arrangement No. 1 should, of course, use a piece of the 35-mm SO-253 film to obtain test-strips with Arrangement No. 2. Regarding troubleshooting, refer to Experimental Arrangement No. 1.

Suggestions for Further Reading:

E.N. Leith, and J. Upatnieks, J Opt Soc Am v53 p1377 (1963)
E.N. Leith, and J. Upatnieks, J Opt Soc Am v54 p1295 (1964)
G.L. Rogers, J Sci Instrum v43 p677 (1966)
J. Landry, J Opt Soc Am v56 p1133 (1966)
L.T. Long, and J.A. Parks, Am J Phys v35 p773 (1967)
G.T. Williams, and T.C. Owen, Phys Educ v2 p278 (1967)
R.H. Webb, Am J Phys v36 p62 (1968)
C.L. Stong, Sci Am p110 (Jly '71)
J. Wise, Am J Phys v40 p1866 (1972)
P.J. Hanford, Phys Educ v8 p380 (1973)
"Introduction to Holography," Instruction Manual 60-688,
    Metrologic Instruments, Inc., Bellmawr, New Jersey 08030 (1975)
"Holography Using a Helium-Neon Laser,"
    Metrologic Instruments, Inc., Bellmawr, New Jersey 08030 (1976)
"Kodak Materials for Holography," Kodak Pamphlet No. p-110 (1976)
J.R. Duckett, G.F. Lothian, and J.F. Parsons, Phys Educ v11 p294 (1976)

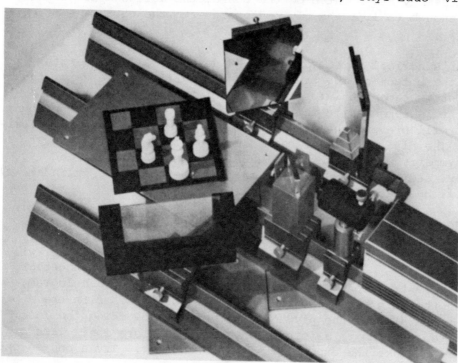

Fig. 257

When viewing planar holograms, it appears that one is looking through a window at the object. Thus, one can only view an object in a 10x12.5 cm (4"x5") hologram as if one were looking through a window of the same size as the film. The advantage of 360-degree holograms is that the 'window' wraps all the way around the object. T.H. Jeong was the first to describe cylindrical holography. Today there are commercial kits available (Metrologic No. 60-634) which not only include the optical parts but also vibration mounts, chemicals and the special 70-mm wide recording film. Figure 258 shows the arrangement for producing 360-degree holograms.

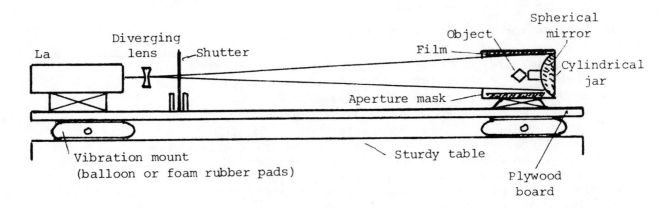

Fig. 258. Arrangement for producing 360-degree holograms.

A heavy plywood board, 30 x 90 cm (1' x 3') makes a good base on which to mount the components. A sturdy table is required underneath the board which is mounted on four inflatable balloons. After the laser and the object/cylinder assembly are mounted on the board, the balloons are lightly inflated, - just enough to lift the plywood board with the equipment off the table.

The diverging spherical mirror, - convex side up - is affixed with rubber cement to the bottom of the transparent cylindrical jar. Next, the little object mount is cemented into the middle of the convex mirror. Finally, the object is cemented to the middle of the top of the object mount. The aperture mask will prevent the diverging laser beam from directly illuminating the film. The diverging laser beam and the cylinder assembly must have a circular symmetry when properly aligned. The object should be illuminated evenly, and there should be no shadows cast onto the cylinder wall where the film will be placed.

A sheet of 70 x 250 mm holographic film is wrapped around the inside of the plastic cylinder. The emulsion side of the film should be toward the object. A few pieces of double-sided adhesive tape inside the cylinder will hold the film and the long ends of the film strip should be taped together. Loading of the film must be done in total darkness. To make the hologram, place the jar with the film enclosed, in its proper position. Lift the shutter (black cardboard) from the plywood base but keep blocking the laser beam for at least ten seconds to eliminate any existing vibration. Then remove the shutter and make the exposure. Replace shutter. Try a one-second exposure initially with a 1 mW laser. For test exposures cut a piece of film into 2-3 cm wide strips and place one of these in the cylinder jar. The film strips should be developed and bleached as previously described.

The developed holograms can be viewed by putting the developed film - emulsion side in but upside down - back into an empty cylindrical jar and placing the jar in the diverged laser beam. A brighter image can be seen by illuminating only a portion of the side of the cylinder. The processed film is formed into a cylinder as it was when exposed and the ends of the strip are held together with tape. The film-cylinder is held in the expanded laser beam with the 'mirror-edge' toward the laser, with the beam striking the emulsion side. Looking through the film-cylinder from the outside an image of the object appears. By rotating the hologram-cylinder one complete revolution a 360-degree view of the object is observed.

References:

T.H. Jeong, P. Rudolph, and A. Luckett, J Opt Soc Am v56 p1263 (1966)
T.H. Jeong, J Opt Soc Am v57 p1396 (1967)
B.A. Stirn, Am J Phys v43 p297 (1975)
W.K. Schubert, and C.R. Throckmorton, Phys Teach p310 (May '75)
Metrologic Instruments, Inc. "Introduction to Holography," p25 (1975)
S.T. Hsue, B.L. Parker, and M. Monahan, Am J Phys v44 p927 (Oct '76)

" " " " " " " " " "

Multiple Recording of Holograms

It has been suggested that many separate holograms could be superimposed on the same piece of film or plate thus making better use of the storage capacity of the holographic emulsion.

Figure 259 is a diagram of the apparatus used for hologram multiplexing. The spatially filtered laser beam is collimated by lens L. B is a beam splitter. M is a front-surface mirror. D is a diffuser. T is the object transparency and H the holographic recording medium.

Multiple recordings can be made:

(a) By varying the angle between the reference beam and the object beam, and hence the spatial frequency of the carrier fringes for each scene recorded. The first exposure is made with the mirror in M position. The second exposure is made with the mirror in M' position. After the exposed film or plate is developed, each scene can be viewed independently by <u>tilting</u> the hologram to different angles relative to the viewing illumination. To avoid 'cross-talk', the reference beam directions (θ, θ', etc.) must be separated sufficiently in angle.

(b) By using the same spatial frequency (mirror M always in the same position) for each exposure but changing the orientation of the carrier fringes on the holographic emulsion. The once exposed holographic film is turned upside down with the emulsion still facing the object side, and a different object transparency is used for the second exposure. Next, the film is turned

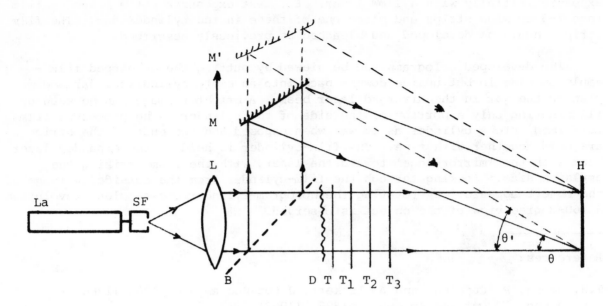

Fig. 259. Arrangement for hologram multiplexing.

90-degrees and a third exposure is made with another object. Finally, the film is turned 180-degrees and exposed with a fourth transparency. The processed hologram has four scenes which can be viewed one at a time by <u>rotating</u> the hologram relative to the viewing illumination.

(c) By using the same spatial frequency for each exposure, as in (b), but by recording the various transparencies in a different plane in space. In other words, first the transparency T is recorded with the system shown in Fig. 259. The transparency T is then removed from the object beam and replaced by transparency T_1 in a different plane in space and recorded on the holographic film. Successively, transparencies T_2, T_3, etc. are recorded on the film. Since each information-bearing transparency is recorded at a different distance (due to the difference in the distance of each transparency from the recording film) they may be separated upon projection by varying the image distance. Such a variation in image distance (projected 'real' images) can be achieved by <u>moving</u> a projection screen from one position to another in space.

Transparent alpha-numeric characters on opaque background make good transparencies for the investigation of the storage capacity of a holographic emulsion. The exposure of each object transparency must be kept to a minimum to avoid saturating the holographic emulsion. The brightness ratio between the reference and object beams should be much lower for multiplexed holograms than the ratio required to produce a satisfactory single hologram.

References:

M. Marchant, and D. Knight, Opt Acta v14 p199 (1967)
R.F. Van Ligten, British Pat. No. 1,106,220 (13 Mch '68)
Battelle Dev. Corp. (Leith-Upatnieks), British Pat. No. 1,104,041 (21 Feb '68),
U.S. Pat. No. 3,506,327 (14 Apr '70)
H.J. Caulfield, Appl Opt v9 p1218 (1970)
P.C. Mehta, and M. Singh, Opt Commun v7 p394 (1973)

Random Bias Holograms

The conventional technique to make holograms of three-dimensional diffuse objects consists of illuminating the high-resolution holographic emulsion with a uniform reference beam which is amplitude modulated by the light reflected from the object. This setup requires beam splitters, lenses, pinhole filters and mirrors.

If, instead of a uniform reference beam, a reference beam is used which has random variations of amplitude from point-to-point, it is still possible to reconstruct a good image from such a hologram. The configuration used for recording random bias holograms is shown in Fig. 260.

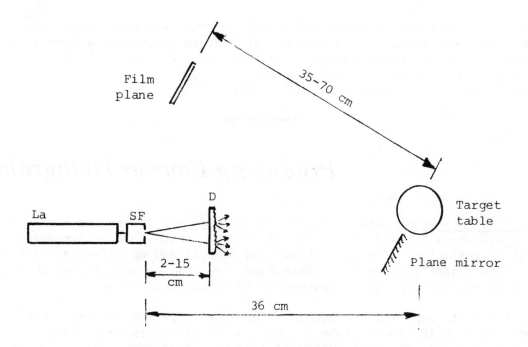

Fig. 260. Recording random bias holograms.

The laser beam impinges directly (*) upon the diffuser which is a piece of ground-glass, non-glare glass, or ordinary tracing paper. The light is scattered from the diffuser D onto the plane front-surface mirror and the three-dimensional object on the target table. The object scatters the light onto the holographic emulsion. The light from the mirror provides a reference beam having a granular structure. The relative intensities of the object and reference beam may be adjusted by rotating the mirror about a vertical axis. *Note: The spatial filter, SF, shown in Fig. 260 is a refinement and is not necessary for the success of the experiment. The diffuser may be attached directly to the exit window of the laser.

After development, the real image reconstructed from the hologram can be photographed by inserting the film directly in the real image plane, or the virtual image may be observed in the usual way by looking through the hologram. Reconstruction is obtained with a point source. Random bias holograms have a speckled appearance but the image quality is good.

While the random bias system leads to some loss of resolution, this is compensated by: (a) system simplicity, (b) shorter set-up time, (c) improved stability, and (d) compact layout.

When using the layout shown in Fig. 260, employing a 1 mW laser, a satin-finish ground-glass 15 cm from the exit window of the laser, and the SO-253 film about 50 cm from the object, the exposure varied from 2 to 10 seconds.

Holographic interferometry is one of the most important applications of holography. The random bias technique was used successfully to make double-exposure holograms of objects under stress.

References:

G.W. Stroke, "An Introduction to Optics of Coherent and Noncoherent Electro-magnetic Radiations," - University of Michigan Engineering Summer Conference on Lasers, Lecture Notes, May 1964, pp1-77
A. Kozma, J Opt Soc Am v56 p428 (1966)
H.H. Arsenault, Opt Commun v4 p267 (1971)

" " " " " " " " " " "

Producing Fourier Holograms

Fresnel holograms refer to a photographic record describing the interference between a Fresnel diffraction pattern (arising from transmission through or reflection from the diffracting object) and an off-axis reference beam. Upon reconstruction these holograms give rise to twin images which are physically separated in space transverse to the optical axis.

A 'perfect' lens takes a bundle of parallel rays and converges them to a single spot on the focal plane. (The Fourier transform of a plane wave is a delta function.) The same 'perfect' lens will transform a point source on the focal plane into a beam of parallel rays. (The Fourier transform of a delta function is a plane wave.)

A Fourier transform hologram is a photographic record of the interference pattern between a Fourier transform of the object amplitude distribution and a coherent background arising from a point reference. Figure 261 shows two systems for creating Fourier transform holograms. Note that since the object appears to be at infinity, the Fourier transform hologram is a Fraunhofer hologram.

Both the transparency (object) and the point source (pinhole) are uniformly illuminated by the same laser source. The system shown in Fig. 261(a) introduces a lens, L, to effect the Fourier transform. The transparency and the point source are in the front focal plane of the lens. The Fourier transform of the point source furnishes the reference beam. The lens also takes the Fourier transform of the light amplitude diffracted by the object transparency. The resultant interference pattern in the back focal plane of the lens is a Fourier transform hologram.

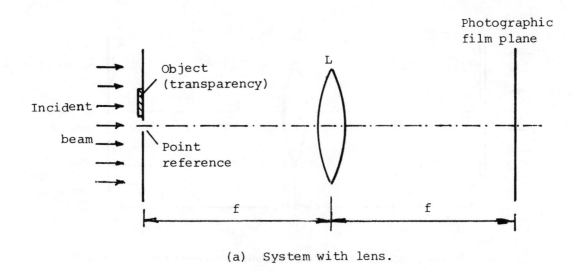

(a)   System with lens.

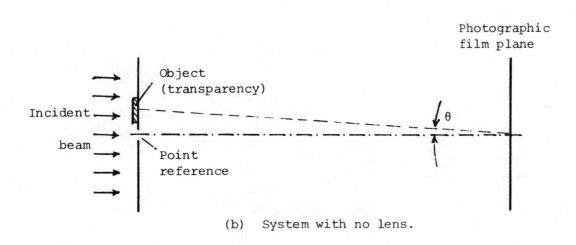

(b)   System with no lens.

Fig. 261.   Formation of Fourier transform holograms.

The second system shown in Fig. 261(b) is the method of "lensless
Fourier transform holography' first described by G.W. Stroke. Even if the
far-field condition is not met in the recording step, the hologram will
reconstruct if the far-field condition is satisfied in the reconstruction
process

The reconstruction from a Fourier transform hologram is done by uniformly
illuminating the hologram as shown in Fig. 262.

In Fourier transform holography, the real and virtual images, reconstructed from the holograms, are formed at +infinity and -infinity, respectively.
The images are simply reconstructed by projecting a collimated laser beam

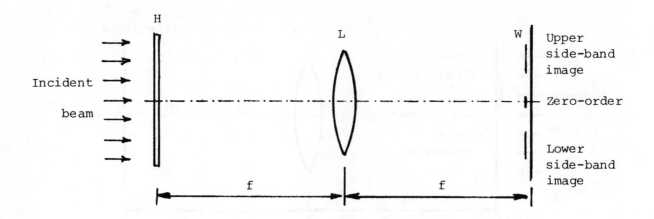

Fig. 262. Fourier transform hologram reconstruction.

through the Fourier transform hologram, H. The images can be photographed in
the focal plane, W, of a lens, L. The lens takes the Fourier transform of the
light diffracted by the hologram; the net effect of two successive Fourier
transforms is to restore the original wavefront. The Fourier transform hologram yields a pair of symmetrical 'real' images (both are 'real', but one
upside down and backward) if the reconstructed wavefront is transformed by
a lens.

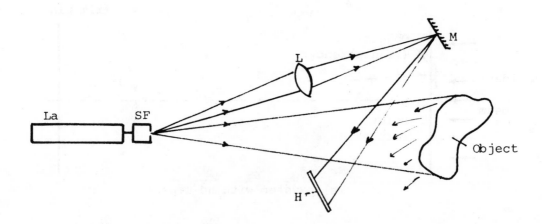

Fig. 263. 'Lensless' Fourier transform hologram recording of 3-D objects.

Figure 263 shows an arrangement for recording holograms of three-dimensional objects with the 'lensless' Fourier transform configuration. The
purpose of lens L is to focus part of the divergent laser beam on the reference mirror M and thus produce a coherent background from a point reference.

By locating the reference mirror on the same optical mount as the object
and the film holder on the same mount as the lens element, a very compact
system is obtained. It can be easily set up on a small optical bench.

Fourier transform holography involves relatively low fringe frequencies
and permits the use of high-speed photographic materials. Good Fourier

transform holographic imaging can be carried out with Polaroid P/N film and on negative materials of similar speed and resolving power.

References:

G.W. Stroke, and D.G. Falconer, Phys Lett v13 p306 (1964)
G.W. Stroke, Appl Phys Lett v6 p201 (1965)
G.W. Stroke, D. Brumm, and A. Funkhouser, J Opt Soc Am v55 p1327 (1965)
G.W. Stroke, A. Funkhouser, C. Leonard, G. Indebetouv, and
      R.G. Zech, J Opt Soc Am v57 p110 (1967)
J.W.C. Gates, and S.J. Bennett, Nature v218 p942 (1968)
S.J. Bennett, and J.W.C. Gates, Nature v221 p1234 (1969)
P. Hariharan, and A. Selvarajan, Opt Commun v4 p392 (1972)
O.A. Shustin, Sov Phys - Uspekhi v14 p668 (1972)

" " " " " " " " " " "

Some Applications of Holography

Classical interferometers can be utilized only for the determination of the quality or mirror-like surfaces, or lenses and polished objects of regular shape. In classical interferometers the wave in one arm is the standard to which the wave in the other arm is compared. Both waves must be present at the same time.

A holographic interferometer does not require specular reflection. It is possible to make interferometric studies of three-dimensional objects of any shape. In holographic interferometry the wave that serves as a standard of comparison can be stored and, if desired, used at a later time. Another important feature is the ability to compare waves that are reflected from a diffusely reflecting object at two different times. Furthermore, a complex object can be examined from many different perspectives, because of the three-dimensional nature of the hologram. If a vibrating surface is investigated, the wavefronts which are reflected by the vibrating surface are time-averaged on the hologram. In the reconstruction, a system of interference bands is produced which determines the nodes and the contours of the spots with a constant vibration amplitude. The single- and double-exposure holographic interferometry makes possible the study of displacement or deformation. The multiple-exposure or time-average technique lands itself to the problem of vibration analysis. A so-called 'live', or 'real-time' (instantaneous) interferometric study of motions and vibrations is also possible, yielding immediate information on the changes.

Holography has widened the scope of interferometry to such a degree that it is fast becoming a standard tool in engineering laboratories all over the world.

Single-exposure Holographic Interferometry

Figure 264 shows a simple holographic interferometer. First, a hologram of a diffusing plate is recorded (Fig. 264-a). After processing, the

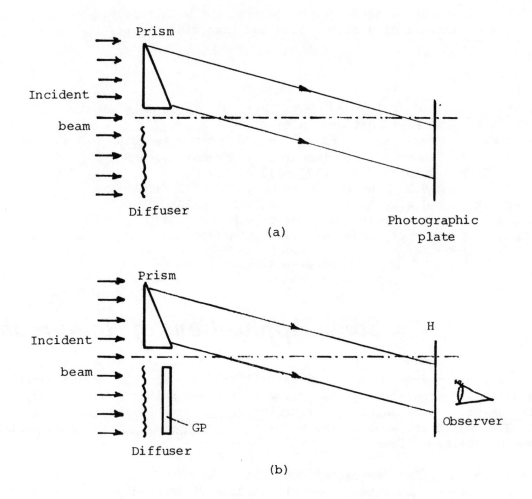

Fig. 264. A simple holographic interferometer.

hologram (H) is placed exactly in its original position (Fig. 264-b). When the hologram is illuminated with the original reference wave, and the diffuser is still in position, the observer will see the reconstructed image of the diffuser in perfect register with the diffuser itself. If a microscope cover glass (GP) is inserted between the diffuser and the hologram (but not intercepting the reference beam) fringes will appear in the image which are related to the wedge and thickness variations and optical inhomogeneities present in the microscope cover glass. The optical quality of GP can be examined from many perspectives in real-time.

A simple experimental configuration for holographic strain interferometry is shown in Fig. 265. The set-up is essentially the same as used in making holograms for any purpose and it is described elsewhere in this book (Experimental Arrangement No. 1). To measure surface deformation at two different times, first a hologram of the object is made without applying the load. The processed hologram is replaced in its exact original position and the load is applied to the object. The observed fringes are related to the magnitude and direction of displacement of every point on the object's surface when the load is applied. This same technique can be used to measure surface deformations of all kinds, including thermal expansion and contractions. For a 'live' fringe study, the hologram must either be developed in place or replaced back in exactly the same position it occupied during its exposure.

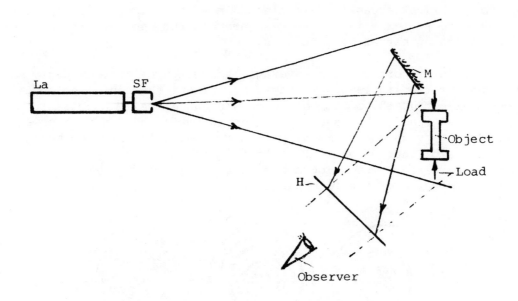

Fig. 265. Holographic strain interferometry.

Of all the methods of holographic interferometry, the single-exposure method is the most difficult to carry out because realignment of the developed hologram is a very critical operation.

Double-exposure Holographic Interferometry

The same setup as shown in Fig. 265 can be used for double-exposure holographic interferometry. With the double-exposure method two holograms are made on a single recording medium. One exposure is made without the load. the load is then applied and another exposure made. With this method the hologram retains, as a permanent record, the change in shape of the object between exposures. During reconstruction both images are seen at once. The two images will interfere with one another and a complete three-dimensional record of the interference phenomena is obtained. With this technique no precision optical components are needed and accurate registration of hologram and object is no longer required. The double-exposure method, on the other hand, is more versatile since the double-exposed hologram can only compare the object and one change in the state of the object. The double-exposure technique is also called 'time-lapse' holographic interferometry.

Time-average Holographic Interferometry

The high-resolution emulsions used in making holograms are of low speed. Exposure times of minutes are not uncommon. Even imperceptible motions of the components forming the optical system may completely blur the recording of the fine fringe pattern. Object undergoing steady state vibration, however, can be recorded. In continuous-exposure hologram interferometry, the object moves continuously during the exposure time. If the motions are very small and sinusoidal, the finished hologram can be thought of as a hologram having been exposed many times. The fringes are contours of constant amplitude. The nodal lines of the vibrating object do not move and will appear as bright areas. The antinodes move throughout the exposure and the interference patterns in these areas are blurred, so that in reconstruction these areas appear as dark fringes.

The earliest report on 'time-average' holographic interferometry was by Powell and Stetson in 1965. In their experiment the vibrating object was a 35-mm film can, the bottom of which was magnetically coupled to a solenoid mounted inside the can. The solenoid was connected to a power-amplifier driven by an audio signal generator. With this arrangement, the frequency and amplitude of excitation to the bottom of the can could be separately controlled. The holographic setup shown in Fig. 265 was used to analyze the vibration of the film can over a selected period of time.

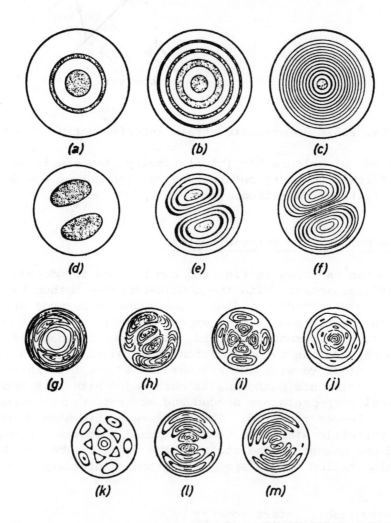

Fig. 266. Mode patterns reconstructed from holograms of a vibrating object.

Figures 266-a through 266-m are replicas of pictures made from hologram reconstructions of the vibrating film can. Figs. 266-a, 266-b and 266-c are the result of the film can vibrating at its lowest frequency of resonance and the differences in the patterns are caused by changes in the amplitude of excitation only. The rings are not mechanical nodes but rather lines characterizing equal amplitudes of vibration. Figs. 266-d, 266-e, and 266-f represent the pattern produced at the second resonant frequency with three different amplitudes. Here the line across the middle is clearly a node of vibration of the can, and the contours to either side are contours of constant amplitude of vibration. Figs. 266-g through 266-m show various other resonant frequencies of the film can bottom as the frequency of excitation was increased.

The above-described method of vibration analysis has widespread applications. The method can be used regardless of the shape or complexity of the object; the vibration modes can be examined from several perspectives; and the analyzed surface need not be optically coated or polished. No mirrors, lines, fibers or sensing mechanisms need to be attached to the surface under analysis. The precision of the measurements may be within a fraction of a micrometer. Any system that operates by mechanical vibrations can be analyzed, such as audio-speaker diaphragms, or musical instruments such as percussion or strings, and audio transducers of many sorts. Also the method is applicable to larger systems, or models of larger systems, such as aerodynamic structures and hydrofoils.

Suggestions for Further Reading:

K.A. Haines, and B.P. Hildebrand, Phys Lett v19 p10 (1965)
R.L. Powell, and K.A. Stetson, J Opt Soc Am v55 p612 (1965)
R.L. Powell, and K.A. Stetson, J Opt Soc Am v55 p1593 (1965)
K.A. Stetson, and R.L. Powell, J Opt Soc Am v55 p1694 (1965)
B.P. Hildebrand, and K.A. Haines, Appl Opt v5 p172 (1966)
K.A. Haines, and B.P. Hildebrand, Appl Opt v5 p595 (1966)
E.N. Leith, and J. Upatnieks, J Opt Soc Am v56 p523 (1966)
J.W. Goodman, W.H. Huntley, Jr., D.W. Jackson, and
    M. Lehmann, Appl Phys Lett v8 p311 (1966)
R.F. van Ligten, and H. Osterberg, Nature v211 p282 (1966)
H.H.M. Chau, and M.H. Horman, Appl Opt v5 p1237 (1966)
E.B. Aleksandrov, and A.M. Bonch-Bruevich, Sov Phys-Tech Phys v12 p258 (1967)
A.F. Metherell, H.M.A. El Sum, J.J. Dreher, and
    L. Larmore, J Acoust Soc Am v42 p735 (1967)
E.N. Leith, and A.L. Ingalls, Appl Opt v7 p539 (1968)
W.G. Gottenberg, Exp Mech v8 p405 (1968)
S. Lu, Proc IEEE v56 p116 (1968)
Battelle Development Corp., British Pat. No. 1,171,445 (18 Nov '69)
J.E. Sollid, Appl Opt v8 p1587 (1969)
R.P. Floyd, and D.J. Collins, Am J Phys v39 p359 (1971)
C.E. Taylor, Am J Phys v39 p417 (1971)
T.D. Dudderar, and R. O'Regan. Materials Res & Standards p8 (Sep '71)
D.H. McMahon, Appl Opt v11 p798 (1972)
R. Levin, El-opt Syst Design p81 (Apr '76)
J.R. Duckett, G.F. Lothian, and J.F. Parsons, Phys Educ v11 p294 (Jne '76)
N. Abramson, Laser Focus v12 p68 (Sep '76)

" " " " " " " " " "

Thin-film Optical Waveguide

Light waves can be guided in thin dielectric films that are the two-dimensional analog of optical fibers. Thin-film optical devices can be made very small and they can be placed next to each other on a single substrate, forming an optical system. In recent years there have been important advances in thin-film light guides. Integrated optical circuits, two-dimensional optical elements, electro-optic, acousto-optic and nonlinear optical devices have been extensively studied.

The prism-film coupler invented by P.K. Tien, R. Ulrich and R.J. Martin of Bell Telephone Laboratories, provides an efficient method of feeding a laser beam into a thin-film optical waveguide. In this coupler (Fig. 267), the prism is placed on top of the thin-film guide. An air gap is maintained between the base of the prism and the top surface of the film. The incident light it totally reflected at the base of the prism, and the evanescent field below the prism then penetrates into the film and excites a light wave in the film. This coupling process is called 'optical tunneling'. For effective coupling, the width of the air gap is less than the wavelength of the laser light.

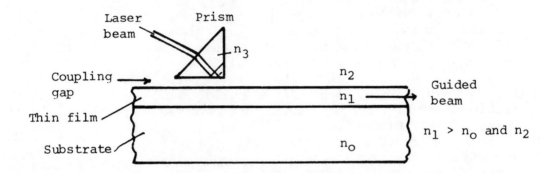

Fig. 267. Arrangement for feeding a laser beam into a thin film.

For a thin film to act as a dielectric waveguide for the light waves, the refractive index of the film n_1 must be larger than that of the substrate n_o and naturally that of the air space above, n_2. Photographic gelatin can be used to make a thin-film waveguide on a glass substrate ($n_o = 1.51$). The refractive index of the gelatin at the operating wavelength ($\lambda = 633$ nm) is about $n_1 = 1.54$. The film is excited by focusing the radiation from a He-Ne laser on it through a 90-degree heavy flint glass prism ($n_3 = 1.65$ to 1.75). For coupling light energy into the film, the right edge of the laser beam is placed as close as possible to the rectangular corner of the prism. When the waveguide is excited by linearly polarized radiation, single-mode operation is possible with film thicknesses from 0.5 to 2.0 μm.

The gelatin light-guiding films can be prepared on glass microscope slides as substrates. The slides must be cleaned with a detergent, rinsed in hot tap water and cold distilled water. While still wet, they are sprayed with methanol and dried. The clean glass plate is immersed in a solution of photographic gelatin in distilled water. The plate is then removed and slowly dried in a vertical position in a clean, dust-free environment. By using sufficiently diluted solutions, films as thin as 0.5 μm can be prepared. For a 7% solution of gelatin the film obtained is approximately 3.0 μm thick.

The prism-film coupler is usually mounted on a turntable so that the laser beam can enter the prism at any angle. The thin film is pressed with adjustable pressure against the base of the prism by a knife edge. The pressure point is about 1 mm or less away from the rectanguler corner of the prism. The dust particles between the prism and the film act as the spacers. The linearly polarized TEM_{00} laser beam is directed on the coupling spot at the prism base. The laser beam is incident upon the prism base at an angle θ. This angle is related to the indices of refraction of the materials involved. Finding the coupling angles requires a certain degree of skill, patience and experience. The light wave induced in the thin film propagates to the right, undergoing multiple reflections inside the film. The multiple interval reflections within the film determine the propagation velocity of the wave in the film according to principles analogous to those applicable to micro-wave waveguides. Coupling can be observed by the appearance of a streak of guided light in the film extending from the coupling spot toward the right. The streak consists of light scattered from the propagating mode, much like a laser beam is visible in dusty air. To show that the light wave is truly propagating inside the film, the continuity of the film is interrupted by scratching across the streak with a fine point. The light streak then ends sharply at the scratched point and the scratched point radiates brightly as an antenna.

Figure 268 shows that the prism-film coupler can be used as an output coupler. Two 90-degree prisms are used, one as an input coupler and the other for output.

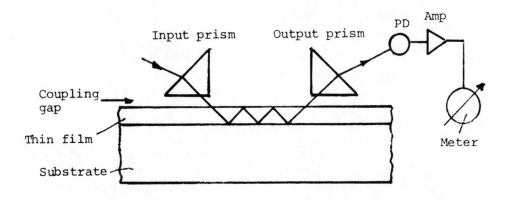

Fig. 268. Measuring the loss in a thin-film guide.

The setup shown in Fig. 268 can be used to measure the loss in a thin film. One prism-film coupler axcites a light streak in the film and a second coupler, some distance from the first, couples the light wave out of the film. The output coupler is applied at different points along the light streak. The light emerging from the output prism is then detected (photo-detector - amplifier - ammeter). The measurements thus obtained at different points along the streak are used to evaluate the loss in the film.

References: (see next page)

References:

P.K. Tien, R. Ulrich, and R.J. Martin, Appl Phys Lett v14 p291 (1 May '69)
P.K. Tien, and R. Ulrich, J Opt Soc Am v60 p1325 (Oct '70)
R. Ulrich, J Opt Soc Am v60 p1337 (Oct '70)
H. Nassenstein, Naturwissenschaften v57 p408 (Oct '70)
D.B. Ostrovsky, and A. Jacques, Appl Phys Lett v18 p556 (15 Jne '71)
R. Ulrich, and R.J. Martin, Appl Opt v10 p2077 (Sep '71)
P.K. Tien, Appl Opt v10 p2395 (Nov '71)
G.R. Wessler, K.H. Roedel, and P. Friese, Exper Tech der Physik v20 p435 ('72)
R. Ulrich, and H.P. Weber, Appl Opt v11 p428 (Feb '72)
L.N. Deryugin, and T.K. Chekhlova, Opt Spectrosc v35 p209 (Aug '73)
R. Ulrich, and R. Torge, Appl Opt v12 p2901 (Dec '73)
V.I. Anikin, and L.N. Deryugin, Opt Spectrosc v39 p546 (Nov '75)
E.M. Zolotov, Sov Phys Quant Electron v6 p247 (Feb '76)

"""""""""""

APPENDICES

Fig. 269. Cylindrical lenses for Lazy Susan exercises.

Fig. 270. Improvised Lazy Susan setup.

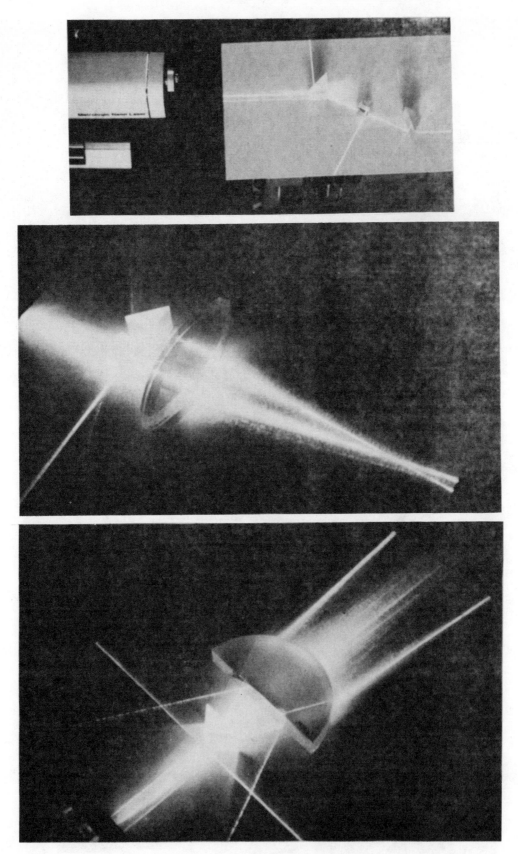

Fig. 271. Some effects observed on the Lazy Susan table.

Fig. 272.   The Michelson interferometer.

Fig. 273.   Voice Communication Set (Metrologic 60-247).

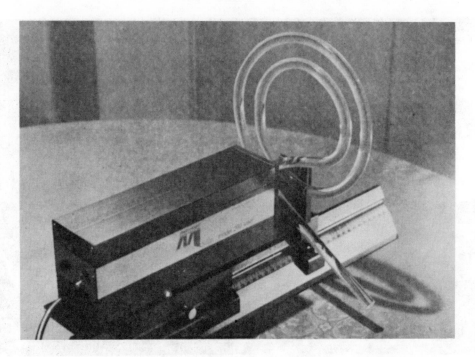

Fig. 274. Laser beam conducted within a 'light pipe'.

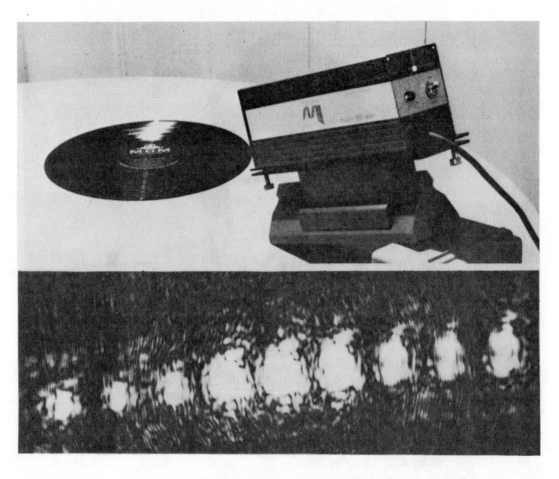

Fig. 275. Long-playing record used as a reflection grating,
and the pattern seen on a nearby wall.

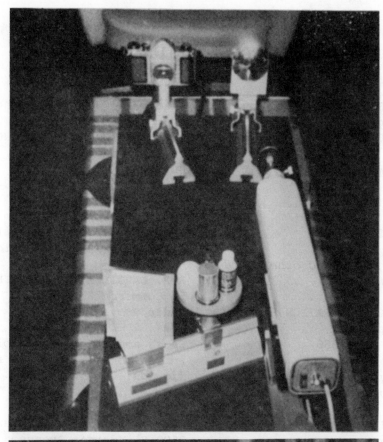

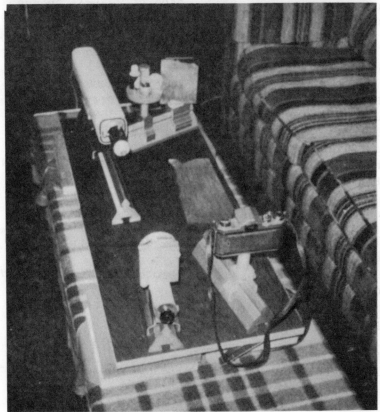

Fig. 276.   Two views of the holographic setup shown in Fig. 254 (p243).

LIST OF SUPPLIERS

ALLIED ELECTRONICS CORPORATION
2400 W. Washington Blvd.
Chicago, IL 60612

Electronics components.

ALTAJ ELECTRONICS
P.O. Box 28592
Dallas, TX 75228

Semiconductors, IC's and other hard-to-get components.

ARDEL KINEMATIC CORPORATION
125-20 18th Avenue
College Point, NY 11356

Translators, optical mounts, gimbal mounts, polar rotators.

AUTOMATION INDUSTRIES, INC.
Sperry Division
Shelter Rock Road
Danbury, Conn. 06810

Ultrasonic transducers.

BDH Chemicals Ltd.
Broom Road, Poole BH12 4NN
E n g l a n d
Agents: Gallard/Schlesinger
        584 Mineola Ave.
        Carle Place, NY 11514

Liquid crystals, single crystals and crystal products (lenses, windows, etc)

BARRY ELECTRONICS
512 Broadway
New York, NY 10012

Electronic components by mail order.

BURLEIGH INSTRUMENTS, INC.
100 Despatch Drive
P.O. Box 388
East Rochester, NY 14445

Fabry-Perots, PZT devices, tunable etalons, ramp generators, optical

erector set, mirror mounts.

BURLEIGH INSTRUMENTS, INC.
Optical Shop Division
Box 862
Upper Montclair, NJ 07043

Interference filters, solid and fixed air gap etalons, Glan-Air polarizers, depolarizers, polarization rotators.

R.P. CARGILLE LABORATORIES, INC.
55 Commerce Road
Cedar Grove, NJ 07009

Epoxy embedding kits, hollow prisms, immersion oil, index of refraction liquids.

CENCO
Central Scientific Company
2600 S. Kostner Avenue
Chicago, IL 60623

Demonstration and lab equipment.

CHANNEL INDUSTRIES, INC.
839 Ward Drive
P.O. Box 3680
Santa Barbara, CA 93105

Piezoelectric ceramic materials.

CONTINENTAL SPECIALTIES CORPORATION
44 Kendall Street
P.O. Box 1942
New Haven, Conn. 06509

"Proto Board" breadboard assemblies.

CRYTON OPTICS, INC.
7 Skillman St.
Roslyn, NY 11576

Fresnel lenses from stock.

DOVER PUBLICATIONS, INC.
180 Varick Street
New York, NY 10014

The "Dover Pictorial Archive Series" is an excellent source of geometrical designs which can be photographically reduced for diffraction experiments.

DUKE STANDARDS COMPANY
445 Sherman Avenue
Palo Alto, CA 94306

Polystyrene micro-spheres for tracing flows and for light scattering experiments.

E.M. LABORATORIES, INC.
500 Executive Blvd.
Elmsford, NY 10523

Complete line of liquid crystals, manufactured by E. Merck, Darmstadt.

THE EALING CORPORATION
Optics Division
22 Pleasant Street
So. Natick, MA 01760

Optical benches, components and accessories, Smartt Point Diffraction Interferometer.

EASTMAN KODAK COMPANY
Eastman Organic Chemicals
Rochester, NY 14650

Liquid crystals and other organic chemicals.

EASTMAN KODAK COMPANY
Industrial Sales
Rochester, NY 14650

Neutral density attenuators.

EASTMAN KODAK COMPANY
Scientific Photography Markets
Professional, Commercial and
Industrial Markets Division
343 State Street
Rochester, NY 14650

Films and plates for holography and for diffraction mask production.

EDMUND SCIENTIFIC CO.
Edscorp Building
Barrington, NJ 08007

Lenses, mirrors, prisms, beam splitters, corner cube reflectors, Lazy Susan turntables, iris diaphragms, Ronchi rulings, diffraction grating replicas, polarizers, optical flats, pinholes, Fresnel lenses, holograms for the experimenter.

EMERSON & CUMING, INC.
Canton, MA 02021

Casting resins, adhesives, ceramics, silicones, hollow glass and plastic spheres, electrically conductive adhesives and coatings. Eastman 910 Adhesive.

GC ELECTRONICS (CALECTRO)
400 S. Wyman St.
Rockford, IL 61101

Semiconductors, hard-to-get IC's and other components.

GAERTNER SCIENTIFIC CORP.
1201 West Wrightwood
Chicago, IL 60014

Optical instruments and lab hardware.

GULTON INDUSTRIES, INC.
Piezo Products Division
P.O. Box 4300
Fullerton, CA 92634

Glennite piezoceramics - HST-41

HEATH/SCHLUMBERGER INSTRUMENTS
Benton Harbor, Michigan 49022

Strip chart recorders, scopes, frequency counters, multimeters, sine/square wave generators.

H.P.M. MARKETING CORP.
98 Commerce Road
Cedar Grove, NJ 07009

Photographic developing tanks and "gepe" glassless slide binders.

INTERNATIONAL PRODUCTS CORP.
P.O. Box 118
Trenton, New Jersey

"Micro" laboratory glassware cleaner.

JENSEN TOOLS
4117 North 44th St.
Phoenix, Arizona 85018

*One-step shopping for quality tools,
off-the-shelf delivery.*

JODON ENGINEERING ASSOCIATES, INC.
145 Enterprise Drive
Ann Arbor, Michigan 48103

*Microscope objectives; OMIT lint and
dust remover; Sticki wax; Epoxy;
Collodion; Reagent grade acetone;
Instruments & systems for optical
data processing, holography and
HNDT; Films and chemicals for holography; Tutorial holograms. Modest
quantity orders accepted.*

KLINGER SCIENTIFIC APPARATUS CORP.
83-45 Parsons Blvd.
Jamaica, NY 11432

*Optical benches, accessories, modular
units, optical components.*

LINDEN LABORATORIES, INC.
Box 920
State College, PA 16801

Piezoceramics and transducers.

MELLES GRIOT
1770 Kettering St.
Irvine, CA 92714

*Lenses, mirrors, beam splitters,
prisms, optical flats, filters,
opto-mechanical components.*

METRIGRAPHICS DIV.
DYNAMICS RESEARCH CORP.
50-60 Concord St.
Wilmington, MA 01887

Resolution test targets.

METROLOGIC INSTRUMENTS, INC.
143 Harding Avenue
Bellmawr, NJ 08030

*Student lasers (He-Ne) and pro lab
lasers, modulatable lasers. Optics
teaching kits. Holography kits.*

*Speed of light kit. Laser power meters.
Photo-detectors. Video receiver.
X-Y position indicator. Collimators.
Spatial filters. Laser communication
kit. Optics bench equipment.*

NATIONAL CAMERA, INC.
2000 West Union Avenue
Englewood, Colorado 80110

*Unusual, hard-to-find tools, products,
and equipment.*

NATIONAL PHOTOCOLOR, CORP.
53 Water Street
South Norwalk, CT 06854

Pellicles.

NEWPORT RESEARCH CORPORATION
18235 Mt. Baldy Circle
Fountain Valley, CA 92708

*Films, plates and chemicals for laser
based photography. Modest quantity
orders accepted. Also table systems,
components, optics and laser related
electronic instruments.*

OPTICAL SCIENCES GROUP, INC.
24 Tiburon Street
San Rafael, CA 94901

*High quality Fresnel lenses and prisms
available from stock.*

ORIEL CORPORATION OF AMERICA
15 Market Street
Stamford, CT 06902

*Optical benches and tables and
accessories. Filters, polarizers,
retardation plates.*

PANAMETRICS
221 Crescent Street
Waltham, MA 02154

Ultrasonic transducers.

PANCRO MIRRORS, INC.
6413 San Fernando Road
Glendale, CA 91201

*Mirror coatings, - will aluminize
glass plates, lenses, etc.*

P.P.G. INDUSTRIES, INC.
One Gateway Center
Pittsburgh, PA 15222

NESA and NESATRON conductive glass.

PRECISION CELLS, INC.
221 Park Avenue
Hicksville, NY 11801

Glass and quartz standard optical cells; micro-cells; cylindrical cells; flow cells; absortiometer, comparator and Raman cells.

RADIO SHACK
2617 West 7th Street
Fort Worth, TX 75107

Electronics components.

SOLID STATE SYSTEMS, INC.
P.O. Box 617
Columbia, MO 65201

Semiconductors, IC's, components.

SPACE OPTICS RESEARCH LABORATORIES
7 Stuart Road
Chelmsford, MA 01824

Fourier transform optics, interferometers, optical instruments.

SPECIAL OPTICS
Box 165
Little Falls, NJ 07424

Optical components for ultraviolet, visible and infrared.

THE L.S. STARRETT COMPANY
Athol, MA 01331

Precision tools. Calipers, micrometers, optical measuring tools.

3M COMPANY
Visual Products, Industrial Optics
3M Center, Bldg. 220-10W
St. Paul, MN 55101

Light control film, lens film, Fresnel lenses, light polarizers, thermofilm, liquid crystals and conductive glass. Lenscreen rear projection screens and light diffusing films.

TELEDYNE GURLEY
514 Fulton Street
Troy, NY 12181

Reticles, resolution targets, step wedges.

VERNITRON PIEZOELECTRIC DIVISION
232 Forbes Road
Bedford, Ohio 44146

Piezoelectric ceramics - PZT.

VINCENT ASSOCIATES
1255 University Avenue
Rochester, NY 14607

Electronically programmable laser beam shutters.

THE WELCH SCIENTIFIC COMPANY
7300 N. Linder Avenue
Skokie, IL 60076

Optical instruments and laboratory hardware.

"""""""""""

BIBLIOGRAPHY

Born, M., and Wolf, E., PRINCIPLES OF OPTICS, (5th ed.) Pergamon Press,
New York, 1975

Crawford, F.S. Jr., WAVES, McGraw-Hill, New York, 1968

Hecht, E., and Zajac, A., OPTICS, Addison-Wesley, Reading, 1973

Hecht, E., THEORY AND PROBLEMS OF OPTICS, McGraw-Hill (Schaum's),
New York, 1975

Jenkins, F.A., and White, H.E., FUNDAMENTALS OF OPTICS, McGraw-Hill,
New York, 1976

Klein, M.V., OPTICS, Wiley, New York, 1970

Lipson, S.G., and Lipson, H., OPTICAL PHYSICS, Cambridge University Press,
London, 1969

Longhurst, R.S., GEOMETRICAL AND PHYSICAL OPTICS, (3rd ed.), Longman Group
Ltd., London, 1973

Meyer-Arendt, J.R., INTRODUCTION TO CLASSICAL AND MODERN OPTICS,
Prentice-Hall, Englewood Cliffs, 1972

Monk, G.S., LIGHT (2nd ed.), Dover Publications, New York, 1963

Palmer, C.H., OPTICS, The Johns Hopkins Press, Baltimore, 1962

Stroke, G.W., AN INTRODUCTION TO COHERENT OPTICS AND HOLOGRAPHY, (2nd ed.)
Academic Press, New York, 1969

Wood, R.W., PHYSICAL OPTICS, (3rd rev. ed.), Dover Publications, N.Y. 1967

"""""""""""

Abeles, F. (ed.), OPTICAL PROPERTIES OF SOLIDS, Elsevier-North Holland,
New York, 1972

Allan, W.B., FIBRE OPTICS, Plenum Press, New York, 1973

Allen, L., and Jones, D.G.C., PRINCIPLES OF GAS LASERS, Plenum Press,
New York, 1967

Andrews, C.L., OPTICS OF THE ELECTROMAGNETIC SPECTRUM, Prentice-Hall,
Englewood Cliffs, 1960

Arecchi, F.T., and Schulz-Dubois, E.O., (eds.), LASER HANDBOOK (2 vols.),
American Elsevier, New York, 1972/76

Arnaud, J.A., BEAM AND FIBER OPTICS, Academic Press, New York, 1976

Baker, B.B., and Copson, E.J., THE MATHEMATICAL THEORY OF HUYGEN'S PRINCIPLE,
Oxford University Press, London, 1969

Baldwin, G.C., AN INTRODUCTION TO NONLINEAR OPTICS, Plenum Press, N.Y., 1974

Ball, C.J., AN INTRODUCTION TO THE THEORY OF DIFFRACTION, Pergamon Press,
New York, 1971

Barber, N.F., EXPERIMENTAL CORRELOGRAMS AND FOURIER TRANSFORMS, Pergamon Press,
New York, 1961

Barnoski, M.K., (ed.), INTRODUCTION TO INTEGRATED OPTICS, Plenum Press,
New York, 1974

Barnoski, M.K., (ed.), FUNDAMENTALS OF OPTICAL FIBER COMMUNICATIONS,
Academic Press, New York, 1976

Barrakette, E.S., et al (eds.), APPLICATIONS OF HOLOGRAPHY, Plenum Press,
New York, 1971

Beesley, M.J., LASERS AND THEIR APPLICATIONS, (2nd ed.), Halsted Press
(Div. of John Wiley), New York, 1976

Beran, M.J., and Parrent, G.B.Jr., THEORY OF PARTIAL COHERENCE, Prentice-Hall,
Englewood Cliffs, 1964

Bindmann, W., DICTIONARY OF OPTICS AND OPTICAL INSTRUMENTS (English-German),
TECHNIK WOERTERBUCH OPTIK UND OPTISCHER GERAETEBAU, (Deutsch-
Englisch), VEB Verlag Technik, Berlin, 1974

Bloembergen, N., NONLINEAR OPTICS, Benjamin, New York, 1965

Bloom, A.L., GAS LASERS, Wiley, New York, 1968

Boutry, G.A., INSTRUMENTAL OPTICS, Hilger and Watts, London, 1961

Bracewell, R.N., THE FOURIER TRANSFORM AND ITS APPLICATIONS, McGraw-Hill,
New York, 1968

Bracey, R.J., THE TECHNIWUE OF OPTICAL INSTRUMENT DESIGN, The English Univ.
Press, London, 1960

Brouwer, W., MATRIX METHODS IN OPTICAL INSTRUMENT DESIGN, Benjamin,
New York, 1964

Brown, E.B., MODERN OPTICS, Reinhold, New York, 1965

Brown, R., LASERS, Doubleday, New York, 1968

Bruhat, G., and Kastler, A., OPTIQUE, (4th ed.), Masson & Cie., Paris, 1954

Cagnet, M., Francon, M., Thrierr, J.C.,ATLAS OF OPTICAL PHENOMENA (1962)
Cagnet, M., Francon, M., Mallick, S., SUPPLEMENT to above volume (1971)
Springer-Verlag, New York

Camatini, E., (ed.), OPTICAL AND ACOUSTICAL HOLOGRAPHY, Plenum Press, New York, 1972

Camatini, F., (ed.), PROGRESS IN ELECTRO-OPTICS, (Vol. 10 in NATO Advanced Study Inst. Ser.) Plenum, N.Y., 1975

Candler, C., MODERN INTERFEROMETERS, Hilger and Watts, London, 1951

Carroll, J.S., PHOTOGRAPHIC LAB HANDBOOK, Amphoto, New York, 1974

Cathey, W.T., OPTICAL INFORMATION PROCESSING AND HOLOGRAPHY, Wiley-Interscience, New York, 1974

Caulfield, H.J., and Lu, S., THE APPLICATIONS OF HOLOGRAPHY, Wiley-Interscience, New York, 1970

Caulfield, H.J., (ed.), COHERENT OPTICAL PROCESSING, Photo-optical Instrumentation Engineers, Palos Verdes Estates, Calif., 1975

Chang, W.S.C., PRINCIPLES OF QUANTUM ELECTRONICS, LASERS: THEORY AND APPLICATIONS, Addison-Wesley Publ. Co., Reading, Mass. 1969

Charschan, S.S. (ed.), LASERS IN INDUSTRY, Van Nostrand Reinhold Co., New York, 1972

Clarke, D., and Grainger, J.P., POLARIZED LIGHT AND OPTICAL MEASUREMENT, Pergamon Press, New York, 1972

Collier, R.J., Burkhardt, C.D., and Lin, L.H., OPTICAL HOLOGRAPHY, Academic Press, New York 1971

Conrady, A.E., APPLIED OPTICS AND OPTICAL DESIGN (Part 1 and 2), Dover Publications, New York, 1957/1960

Cook, A.H., INTERFERENCE OF ELECTROMAGNETIC WAVES, Clarendon, Oxford, 1971

Cowley, J.M., DIFFRACTION PHYSICS, American Elsevier, New York, 1975

Cox, A., PHOTOGRAPHIC OPTICS, Focal Press, London, 1966

Curry, C., WAVE OPTICS - INTERFERENCE AND DIFFRACTION, E. Arnold, London, 1957

Dainty, J.C. (ed.), TOPICS IN APPLIED PHYSICS, VOL.9: LASER SPECKLE AND RELATED PHENOMENA, Springer-Verlag, New York, 1975

Debrus, S., Francon, M., and May, M., OPTICAL INSTRUMENT TECHNIQUES, Oriel Press, Newcastle upon Tyne, 1969

De Gennes, P.G., THE PHYSICS OF LIQUID CRYSTALS, Oxford Univ. Press, N.Y. 1974

Deve, C., OPTICAL WORKSHOP PRINCIPLES (2nd ed.), Hilger & Watts, London, 1954

DeVelis, J.B., Reynolds, G.O., THEORY AND APPLICATIONS OF HOLOGRAPHY, Addison-Wesley, Reading, Mass. 1967

Dickson, J.H., OPTICAL INSTRUMENTS AND TECHNIQUES, Oriel Press, Newcastle upon Tynes, 1970

Ditchburn, R.W., LIGHT (3rd Ed.), Academic Press, New York, 1976

Drude, P., THEORY OF OPTICS, Diver Publications, New York, 1959

Duffieux, P.M., L'INTEGRALE DE FOURIER ET SES APPLICATIONS A L'OPTIQUE, Faculte des Sciences, Besancon, 1946

Durst, F., Melling, A., and Whitelaw, J.H., PRINCIPLES AND PRACTICE OF LASER-DOPPLER ANEMOMETRY, Academic Press, New York, 1976

Eaglesfield, C.C., LASER LIGHT: FUNDAMENTALS AND OPTICAL COMMUNICATION, St. Martin's Press, New York, 1967

Efron, A., EXPLORING LIGHT, Hayden Book Co., New York, 1969

Elion, H.A., LASER SYSTEMS AND APPLICATIONS, Pergamon Press, New York, 1967

Elmer, W.B., THE OPTICAL DESIGN OF REFLECTORS, W.B. Elmer, Andover, 1974

Erf, R.K. (ed.), HOLOGRAPHIC NONDESTRUCTIVE TESTING, Academic, New York, 1974

Essen, L., and Froome, K.D., THE VELOCITY OF LIGHT, Academic Press, N.Y., 1969

Farhat, N.H. (ed.), ADVANCES IN HOLOGRAPHY, (Vol.1) Marçel Dekker, N.Y. 1975

Fishlock, D. (ed.), A GUIDE TO THE LASER, American Elsevier, New York, 1967

Fowles, G.R., INTRODUCTION TO MODERN OPTICS,(2nd ed.), Holt, Reinhart and Winston, Inc., New York, 1975

Francon, M., MODERN APPLICATIONS OF PHYSICAL OPTICS, Wiley, New York, 1963

Francon, M., DIFFRACTION, Gauthier-Vilars, Paris, 1964

Francon, M., OPTICAL INTERFEROMETRY, Academic Press, New York, 1966

Francon, M., DIFFRACTION: COHERENCE IN OPTICS, Pergamon Press, Oxford, 1966

Francon, M., Krauzman, N., Mathieu, J.P., and May, M., EXPERIMENTS IN PHYSICAL OPTICS, Gordon and Breach Science Publishers, New York, 1970

Francon, M., Mallick, S., POLARIZATION INTERFEROMETERS, Wiley, New York, 1971

Francon, M., HOLOGRAPHY, Academic Press, New York, 1974

French, A.P., VIBRATIONS AND WAVES, Norton, New York, 1971

Frocht, M.M, PHOTOELASTICITY, Wiley, New York, 1948

Fruengel, F., HIGH SPEED PULSE TECHNOLOGY - Vol.2 - OPTICAL PULSES, LASERS, MEASURING TECHNIQUES, Academic Press, New York, 1965

Fry, G.A., GEOMETRICAL OPTICS, Chilton, Philadelphia, 1969

Gagliardi, R.M., and Karp, S., OPTICAL COMMUNICATIONS, Wiley, New York, 1976

Garbuny, M., OPTICAL PHYSICS, Academic Press, New York, 1965

Garrett, C.B.G., GAS LASERS, McGraw-Hill, New York, 1967

Gerrard, A., and Burch, J.M., INTRODUCTION TO MATRIX METHODS IN OPTICS,
          Wiley, New York, 1975

Ghatak, A.K., AN INTRODUCTION TO MODERN OPTICS, McGraw-Hill, New York, 1972

Gill, T.P., THE DOPPLER EFFECT, Academic Press, New York, 1965

Goldman, L., BIOMEDICAL ASPECTS OF THE LASER, Springer-Verlag, New York, 1967

Goldman, L., APPLICATIONS OF THE LASER, CRC Press, Cleveland, 1973

Goldwasser, E.L., OPTICS, WAVES, ATOMS, AND NUCLEI, Benjamin, New York, 1965

Goodman, J.W., INTRODUCTION TO FOURIER OPTICS, McGraw-Hill, New York, 1968

Gottenberg, W.G. (ed.), APPLICATIONS OF HOLOGRAPHY IN MECHANICS,
     American Society of Mechanical Engineers, New York, 1971

Gray, G.W., MOLECULAR STRUCTURE AND THE PROPERTIES OF LIQUID CRYSTALS,
                    Academic Press, New York, 1962

Greguss, P. (ed.), HOLOGRAPHY IN MEDICINE, IPC Science and Technology Press,
                    Guildford, Surrey, England, 1976

Grivet, P., ELECTRON OPTICS, Pergamon Press, Oxford, 1965

Guild, J., INTERFERENCE SYSTEMS OF CROSSED DIFFRACTION GRATINGS, Oxford
                    University Press, Oxford, 1956

Guild, J., DIFFRACTION GRATINGS AS MEASURING SCALES, Oxford University
                    Press, Oxford, 1960

Habell, K.J., and Cox, A., ENGINEERING OPTICS, Pitman, London, 1958

Harburn, G., Taylor, C.A., and Welberg, T.R., ATLAS OF OPTICAL TRANSFORMS,
     Cornell University Press, Ithaca and G. Bell & Sons, London, 1975

Hardy, A.C., and Perrin, F.H., THE PRINCIPLES OF OPTICS, McGraw-Hill, N.Y. 1932

Harry, J.E., INDUSTRIAL LASERS AND THEIR APPLICATIONS, McGraw-Hill, N.Y. 1974

Harvey, A.F., COHERENT LIGHT, Wiley-Interscience, New York, 1971

Heavens, O.S., THIN FILM PHYSICS, Barnes & Noble, New York, 1970

Heavens, O.S., LASERS, Scribner's, New York, 1973

Herzberger, M., MODERN GEOMETRICAL OPTICS, Wiley-Interscience, New York, 1958

Hilton, W.A., EXPERIMENTS IN OPTICAL PHYSICS, (3rd ed.), Pillsbury Dept. of
     Physics, William Jewell College, Liberty, Missouri, 1974

Hodam, F., FORMELSAMMLUNG UND TABELLENBUCH DER TECHNISCHEN OPTIK,
                              VEB Verlag, Technik, Berlin, 1974

Holland, L., VACUUM DEPOSITION OF THIN FILMS, Chapman & Hall, London, 1961

Horne, D.F., OPTICAL PRODUCTION TECHNOLOGY, Crane, Russak & Co., Inc.,
                              New York, 1972

Horne, D.F., DIVIDING, RULING AND MASK-MAKING, Crane, Russak & Co., Inc.,
                              New York, 1974

Horne, D.F., LENS MECHANISM TECHNOLOGY, Crane, Russak & Co. Inc.,
                              New York, 1976

Houston, J.B. Jr., OPTICAL SHOP NOTEBOOK (Vol. 1, 1974-75), Optical Society
                              of America, Washington, DC 1975

Huang, T.S., PICTURE PROCESSING AND DIGITAL FILTERING (vol. 6 in "Topics in
     Applied Physics"), Springer-Verlag, New York, 1975

Hyzer, W.G., ENGINEERING AND SCIENTIFIC PHOTOGRAPHY, Macmillan, New York, 1962

Ingalls, A.G. (ed.), AMATEUR TELESCOPE MAKING, (Vols. 1-3),
                              Scientific American, New York, 1953

Jennison, R.C., FOURIER TRANSFORMS AND CONVOLUTIONS FOR THE EXPERIMENTALIST,
     Pergamon Press, New York, 1961

Jeong, T.H., HOLOGRAPHY MANUAL, Gaertner Scientific Corp., Chicago, 1968

Jeong, T.H., HOLOGRAPHY USING A HELIUM-NEON LASER, Metrologic Instruments,
                              Bellmawr, New Jersey, 1976

Joseph, A., and Leahy, D.J., Programmed Physics, Pt. III: OPTICS AND WAVES,
                              Wiley, New York, 1968

Kallard, T. (ed.), HOLOGRAPHY...1969 (Patents & bibliography), 1969
                   HOLOGRAPHY...1970   "   "   1970
                   HOLOGRAPHY...1971/72   "   "   1972
                   LIQUID CRYSTALS AND THEIR APPLICATIONS,   1970
                   ACOUSTIC SURFACE WAVES AND ACOUSTO-OPTIC DEVICES, 1971
                   LIQUID CRYSTAL DEVICES, 1973
                              Optosonic Press, New York, N.Y.

Kaminow, I.P., AN INTRODUCTION TO ELECTRO-OPTIC DEVICES,
                              Academic Press, New York, 1974

Kapany, N.S., FIBER OPTICS, PRINCIPLES AND APPLICATIONS,
                              Academic Press, New York, 1967

Kiemle, H., and Roess, D., EINFUEHRUNG IN DIE TECHNIK DER HOLOGRAPHIE,
                              Akad. Verlagsgesellschaft, Frankfurt,

Plenum Press, New York, 1971

Kingslake, R. (ed.), APPLIED OPTICS AND OPTICAL ENGINEERING, (vols. 1-5),
Academic Press, New York, 1965-1970

Kissam, P., OPTICAL TOOLING FOR PRECISE MANUFACTURE AND ALIGNMENT,
McGraw-Hill, New York, 1962

Klauder, J.R., and Sudarshan, E.C.G., FUNDAMENTALS OF QUANTUM OPTICS,
Benjamin, New York, 1968

Klein, H.A., HOLOGRAPHY: WITH AN INTRODUCTION TO THE OPTICS OF DIFFRACTION,
INTERFERENCE AND PHASE DIFFERENCES, J.B. Lippincott Company, 1970

Kline, M., and Kay, I.W., ELECTROMAGNETIC THEORY AND GEOMETRICAL OPTICS,
Interscience Publ. New York, 1965

Knittl, Z., OPTICS OF THIN FILMS, Wiley, New York, 1976

Kock, W.E., SOUND WAVES AND LIGHT WAVES, Doubleday/Anchor, Garden City, 1965

Kock, W.E., LASERS AND HOLOGRAPHY, Doubleday/Anchor, Garden City, 1969

Kock, W.E., ENGINEERING APPLICATIONS OF LASERS AND HOLOGRAPHY,
Plenum Press, New York, 1975

Koenig, A.D., und Koehler, H., DIE FERNROHRE UND ENTFERNUNGSMESSER (3rd ed.),
Springer-Verlag, New York, 1959

Langenbeck, P., OPTICAL INSTRUMENTS AND TECHNIQUES, Oriel Press, Newcastle
upon Tyne, 1969

Lehmann, M., HOLOGRAPHY - TECHNIQUE AND PRACTICE, Focal Press, London, and
Hastings House, New York, 1970

Lengyel, B.A., INTRODUCTION TO LASER PHYSICS, Wiley, New York, 1966

Lengyel, B.A., LASERS (2nd ed.), Wiley, New York, 1971

Levi, L., APPLIED OPTICS: A GUIDE TO OPTICAL SYSTEM DESIGN, Wiley, New York,'68

Levi, L., HANDBOOK OF TABLES OF FUNCTIONS FOR APPLIED OPTICS,
CRC Press, Cleveland, 1974

Levine, A.K. (ed.), LASERS: A SERIES OF ADVANCES (vols. 1-4), Marcel Dekker,
New York, 1966-1976

Lighthill, M.J., INTRODUCTION TO FOURIER ANALYSIS AND GENERALIZED FUNCTIONS,
Cambridge University Press, 1960

Linfoot, E.H., RECENT ADVANCES IN OPTICS, Clarendon Press, Oxford, 1955

Linfoot, E.H., OPTICAL IMAGE EVALUATION, Focal Press, London, 1960

Linfoot, E.H., FOURIER METHODS IN OPTICAL IMAGE EVALUATION, Focal Press,
London, 1964

Lipson, H.(ed.), OPTICAL TRANSFORMS, Academic Press, New York, 1972

Lloyd, J.M., THERMAL IMAGING SYSTEMS, Plenum Press, New York, 1975

Lothian, G.F., OPTICS AND ITS USES, Van Nostrand Reinhold, New York, 1975

Loudon, R., THE QUANTUM THEORY OF LIGHT, Oxford Univ. Press, New York, 1973

Maitland, A., and Dunn, M.H., LASER PHYSICS (2nd ed.), North-Holland, and
American Elsevier, New York, 1976

Mandel, L., and Wolf, E. (eds.), COHERENCE AND QUANTUM OPTICS,
Plenum Press, New York, 1973

Marshall, S.L. (ed.), LASER TECHNOLOGY AND APPLICATIONS, McGraw-Hill,
New York, 1969

Martin, L.C., and Welford, W.T., TECHNICAL OPTICS (2 vols.),
Pitman, London, 1966

Mathieu, J.P., OPTICS (Parts, 1 and 2), Pergamon Press, New York, 1975

Meier, G., Sackmann, E., and Grabmaier, J.G., APPLICATIONS OF LIQUID CRYSTALS,
Springer-Verlag, New York, 1975

Meiners, H.F. (ed.), PHYSICS DEMONSTRATION EXPERIMENTS (2 vols.),
The Ronald Press Co., New York, 1970

Mertz, L., TRANSFORMATIONS IN OPTICS, Wiley, New York, 1965

Merzkirch, W., FLOW VISUALIZATION, Academic Press, New York, 1974

Metherell, A.F., et al (eds.), ACOUSTICAL HOLOGRAPHY (vols. 1-6),
Plenum Press, New York, 1969-1975

Michelson, A.A., LIGHT WAVES AND THEIR USES, University of Chicago Press,
Chicago, 1961

Michelson, A.A., STUDIES IN OPTICS, University of Chicago Press, Chicago, 1962

Minnaert, M., THE NATURE OF LIGHT AND COLOUR IN THE OPEN AIR, Dover, N.Y.1959

Mollet, P. (ed.), OPTICS IN METROLOGY, Pergamon Press, 1960

Murray, R.D. (ed.), APPLICATIONS OF LASERS TO PHOTOGRAPHY AND INFORMATION
HANDLING, Society of Photographic Scientists and Engineers,
Washington, D.C., 1968

Neblette, C.B. (ed.), PHOTOGRAPHY: ITS MATERIALS AND PROCESSES (7th ed.),
Van Nostrand, Princeton, 1976

Náray, Zs., LASER UND IHRE ANWENDUNGEN, Akadémiai Kiadó, Budapest, Hungary

Nesterikhin, Yu.E., Stroke, G.W., and Kock, W.E. (eds.), OPTICAL INFORMATION
PROCESSING, Plenum Press, New York, 1976

Nussbaum, A., GEOMETRIC OPTICS: AN INTRODUCTION, Addison-Wesley, 1968

Nussbaum, A., and Phillips, R.A., CONTEMPORARY OPTICS FOR SCIENTISTS AND
    ENGINEERS, Prentice-Hall, Englewood Cliffs, New Jersey, 1976

O'Neill, E.L., INTRODUCTION TO STATISTICAL OPTICS, Addison-Wesley, 1963

Orszag, A., LES LASERS - PRINCIPES, REALISATIONS, APPLICATIONS,
                              Masson et Cie., Paris, 1968

Papoulis, A., THE FOURIER INTEGRAL AND ITS APPLICATIONS, McGraw-Hill,
                              New York, 1962

Papoulis, A., SYSTEMS AND TRANSFORMS WITH APPLICATIONS IN OPTICS, McGraw-Hill,
                              New York, 1968

Parrent, G.B.Jr., and Thompson, B.J., PHYSICAL OPTICS NOTEBOOK, Society of
    Photo-optical Instrumentation Engineers, Palos Verdes Estates, CA, 1969

Pearson, J.M., A THEORY OF WAVES, Allyn and Bacon, Boston, 1966

Pohl, R.W., OPTIK UND ATOMPHYSIK (12th ed.), Springer-Verlag, 1967

Pressley, R.J. (ed.), C.R.C. HANDBOOK OF LASERS WITH SELECTED DATA ON OPTICAL
    TECHNOLOGY, CRC Press, Cleveland, 1971

Pratt, W.K., LASER COMMUNICATION SYSTEMS, Wiley, New York, 1969

Preston, K. Jr., COHERENT OPTICAL COMPUTERS, McGraw-Hill, New York, 1972

Ratner, A.M., SPECTRAL, SPATIAL AND TEMPORAL PROPERTIES OF LASERS,
                          Plenum Press, New York, 1975

Robertson, E.R., Harvey, J.M. (eds.), THE ENGINEERING USES OF HOLOGRAPHY,
                          Cambridge Univ. Press, London, 1970

Roess, D., LASERS - LIGHT AMPLIFIERS AND OSCILLATORS, Academic Press,
                            New York, 1969

Rogers, G.L., HANDBOOK OF GAS LASER EXPERIMENTS, Iliffe Books, London, 1970

Ronchi, V., THE NATURE OF LIGHT, Harvard University Press, Cambridge, 1971

Rosenberger, D., et al, TECHNISCHE ANWENDUNGEN DES LASERS,
                       Springer-Verlag, New York, 1975

Ross, M., LASER RECEIVERS, DEVICES, TECHNIQUES, SYSTEMS, Wiley, New York, 1966

Ross, M., LASER APPLICATIONS (vols. 1 & 2), Academic Press, New York, 1971/74

Rossi, B., OPTICS, Addison-Wesley, Reading, Massachusetts, 1957

Rousseau, M., and Mathieu, J.P., PROBLEMS IN OPTICS, Pergamon, Elmsford,
                          New York, 1973

Saltonstall, R., et al, THE COMMERCIAL DEVELOPMENT AND APPLICATION OF
    LASER TECHNOLOGY, Hobbs, Dorman, New York, 1965

Sanders, J.H., DIE LICHTGESCHWINDIGKEIT, Akademie-Verlag, Berlin, 1970

Sapriel, J., L'ACOUSTO-OPTIQUE, Masson et Cie., Paris, 1976

Sargent III, M., Scully, M.O., and Lamb, W.E.Jr., LASER PHYSICS,
                            Addison-Wesley, Reading, 1974

Schawlow, A.L.,(intro.), LASERS AND LIGHT; READINGS FROM SCIENTIFIC AMERICAN,
                        Freeman, San Francisco, 1969

Schulz, G., PARADOXA AUS DER OPTIK, J.A. Barth, Leipzig, 1974

Schwarz, H.J., and Hora, H. (eds.), LASER INTERACTION AND RELATED PLASMA
    PHENOMENA, Plenum Press, New York, 1971

Shulman, A.R., OPTICAL DATA PROCESSING, Wiley, New York, 1970

Shurcliff, W.A., POLARIZED LIGHT: PRODUCTION AND USE, Harvard University Press,
                            Cambridge, 1962

Shurcliff, W.A., Ballard, S.S., POLARIZED LIGHT, Van Nostrand, Princeton, 1964

Siegman, A.E., AN INTRODUCTION TO LASERS AND MASERS, McGraw-Hill, N.Y. 1971

Sinclair, D.C. and Bell, W.E., GAS LASER TECHNOLOGY, Holt, Reinhart and
                        Winston, New York, 1969

Smith, F.G., and Thomson, J.H., OPTICS, Wiley, New York, 1971

Smith, H.M., PRINCIPLES OF HOLOGRAPHY (2nd ed.), Wiley, New York, 1975

Smith, W.J., MODERN OPTICAL ENGINEERING, McGraw-Hill, New York, 1966

Smith, W.V., and Sorokin, P.P., THE LASER, McGraw-Hill, New York, 1966

Smith, W.V., LASER APPLICATIONS, Artech House, Dedham, Massachusetts, 1970

Sommerfeld, A., OPTICS, LECTURES ON THEORETICAL PHYSICS, Vol. 4,
                        Academic Press, New York,(1954), 1964

Southall, J.P.C., MIRRORS, PRISMS, AND LENSES (3rd ed.), Macmillan Co.,
                        New York, 1936

Southall, J.P.C., INTRODUCTION TO PHYSIOLOGICAL OPTICS, Dover, New York, 1961

Steels, W.H., INTERFEROMETRY, University Press, Cambridge, 1967

Steele, E.L., OPTICAL LASERS IN ELECTRONICS, Wiley, New York, 1968

Stimson, A., PHOTOMETRY AND RADIOMETRY FOR ENGINEERS, Wiley, New York, 1974

Stone, J.M., RADIATION AND OPTICS, McGraw-Hill, New York, 1963

Street, C., GRAPHICAL RAY TRACING, Society of Photo-optical Instrumentation
        Engineers, Palos Verdes Estates, California, 1974

Stroke, G.W., er al (eds.), ULTRASONIC IMAGING AND HOLOGRAPHY,
                        Plenum Press, New York, 1974

Strong, J., CONCEPTS OF CLASSICAL OPTICS, Freeman, San Francisco, 1958

Summer, W., PHYSICAL LABORATORY HANDBOOK, Van Nostrand, Princeton, 1966

Svelto, O., PRINCIPI DEL LASER, Tamburini Editore, Milano, 1970 and
                        Plenum Press, New York, 1976

Swindell, W. (ed.), POLARIZED LIGHT, Halsted, New York, 1975

Tamir, T. (ed.), TOPICS IN APPLIED PHYSICS, VOL. 7: INTEGRATED OPTICS,
                        Springer-Verlag, New York, 1975

Taylor, C.A., and Lipson, H., OPTICAL TRANSFORMS, Bell, London, 1964

Thomas, W. Jr. (ed.), S.P.S.E. HANDBOOK OF PHOTOGRAPHIC SCIENCE AND ENGINEERING,
                        Wiley, New York, 1973

Tippett, J.T. et al (eds.), OPTICAL AND ELECTRO-OPTICAL INFORMATION PROCESSING,
                        MIT Press, Cambridge, Mass., 1965

Tolansky, S., SURFACE MICROPHOTOGRAPHY, Wiley, New York, 1962

Tolansky, S., CURIOSITIES OF LIGHT RAYS AND LIGHT WAVES,
                        American Elsevier, New York, 1965

Tolansky, S., REVOLUTION IN OPTICS, Penguin Books, Baltimore, 1968

Tolansky, S., MULTIPLE-BEAM INTERFEROMETRY OF SURFACES AND FILMS,
                        Dover Publications, New York, 1970

Tolansky, S., AN INTRODUCTION TO INTERFEROMETRY (2nd ed.), Wiley,
                        New York, 1973

Tomiyasu, K., THE LASER LITERATURE: AN ANNOTATED GUIDE, Plenum Press, NY 1968

Towne, D.H., WAVE PHENOMENA, Addison-Wesley, Reading, Massachusetts, 1967

Tradowky, K., LASER - KURZ UND BUENDIG, Vogel-Verlag, Wuerzburg, 1968

Troup, C.J., OPTICAL COHERENCE THEORY - RECENT DEVELOPMENTS,
                        Methuen and Co., London, 1967

Twyman, F., PRISM AND LENS MAKING, Hilger & Watts, London, 1957

Valasek, J., OPTICS, THEORETICAL AND EXPERIMENTAL, Wiley, New York, 1949

Van de Hulst. H.C., LIGHT SCATTERING BY SMALL PARTICLES,
                        Wiley, New York, 1957

Van Heel, A.C.S. (ed.), ADVANCED OPTICAL TECHNIQUES, Wiley, New York, 1967

Van Heel, A.C.S., and Velzel, C.H.F., WHAT IS LIGHT, McGraw-Hill, New York,'68

Vasiček, A., OPTICS OF THIN FILMS, Wiley, New York, 1960

Wagner, A.F., EXPERIMENTAL OPTICS, Wiley, New York. 1929

Wahlstrom, E.E., OPTICAL CRYSTALLOGRAPHY (4th ed.), Wiley, New York, 1969

Wall, E.J., and Jordan, F.I., PHOTOGRAPHIC FACTS AND FORMULAS,
                              Amphoto, New York, 1974

Walsh, J.W.T., PHOTOMETRY, Dover Publications, New York, 1965

Watrasiewicz, B.M., and Rudd, M.J., LASER DOPPLER MEASURMENTS,
                              Butterworths & Co., Boston, 1976

Webb, R.H., ELEMENTARY WAVE OPTICS, Academic Press, New York, 1969

Weber, H., and Herziger, G., LASER, GRUNDLAGEN UND ANWENDUNGEN,
                              Physik-Verlag, Weinheim, 1972

Welford, W.T., OPTICS, (Vol. 14, Oxford Physics Series), Oxford University
                              Press, London, 1976

Williams, C.S., and Becklund, D.A., OPTICS: A SHORT COURSE FOR ENGINEERS
    AND SCIENTISTS, Wiley-Interscience, New York, 1972

Wolf, E. (ed.), PROGRESS IN OPTICS (vols. 1-14), North-Holland Publishing Co.,
    Amsterdam and American Elsevier Publishing Co., New York, 1961-1976

Wood, E.A., CRYSTALS AND LIGHT, Van Nostrand Co., Princeton, 1964

Wright, G., ELEMENTARY EXPERIMENTS WITH LASERS, (Wykeham Sci.Ser., vol. 18),
                              Springer-Verlag, New York, 1973

Yariv, A., QUANTUM ELECTRONICS (2nd ed.), Wiley, New York, 1975

Young, H.D., FUNDAMENTALS OF WAVES, OPTICS, AND MODERN PHYSICS, (2nd ed.),
                              McGraw-Hill, New York, 1976

Yu, F.T.S., INTRODUCTION TO DIFFRACTION, INFORMATION PROCESSING AND
    HOLOGRAPHY, MIT Press, Cambridge, Mass., 1973

Yu, F.T.S., OPTICS AND INFORMATION PROCESSING, Wiley, New York, 1976

Zernike, F., and Midwinter, J.E., APPLIED NON-LINEAR OPTICS,
                              Wiley, New York, 1973

Zimmer, H.G., GEOMETRICAL OPTICS, Springer-Verlag, New York, 1970

""""""""""

PRIMARY SOURCES OF INFORMATION:

These publications contain chiefly new material or new presentations and discussions of known material. In general, they contain the latest published information. Examples are periodicals, governmental reports, patents, and manufacturers' technical pamphlets.

The optical and electro-optical trade magazines are probably the best way to keep up with advances in the field: *Electro-optical Systems Design*, *Laser Focus*, and *Optical Spectra*. *Laser Focus* regularly publishes abstracts of contemporary laser related papers that have appeared in other journals or as a governmental report.

Lists of companies selling, leasing, and renting laser systems, services, or equipment can be obtained from the latest edition of: *Laser Focus* "Annual Buyers Guide," and *Electro-optical Systems Design* "Vendor Selection Issue."

The most important journals in which articles on laser applications appear regularly are, in alphabetical order:

American Journal of Physics
Applied Optics
Applied Physics Letters
Comptes Rendus (Ser. B)
Experimental Mechanics
IEEE Proceedings
Japanese Journal of Applied Physics
Journal of Applied Photographic Engineering (SPSE)
Journal of the Optical Society of America
Nature
Nouvelle Revue d'Optique
Optica Acta
Optical Engineering (SPIE)
Optical Sciences Newsletter (University of Arizona)
Optics and Laser Technology
Optics and Spectroscopy
Optics Communications
Optik
Photographic Science and Engineering (SPSE)
Physics Education
Physics Teacher, The
Review of Scientific Instruments
Scientific American
Scientific Instruments (Br. Journal of Physics, Pt.E)
Soviet Journal - Optical Technology
Soviet Journal - Quantum Electronics
Soviet Physics - Technical Physics
Soviet Physics - Uspekhi

Patents are a good source of technical information since they usually present detailed explanations of how the patented device or process works. The journal *Applied Optics* has been reviewing optics patents for a number of years. Copies of U.S. Patents may be obtained from: The Commissioner of Patents, Washington, D.C. 20231 - for 50¢ per copy, paid in advance. The *Official U.S. Patent Office Gazette* is a weekly publication that lists patents as they are issued.

INDEX

Mustacich, R.V.,   179
Myers, G.E.,   8
Myers, O.E.,   236

Naba, N.,   56
Nakamura, K.,   206
Nassenstein, H.,   262
Neeson, J.F.,   212
Nefarrate, A.B.,   204
Nicklin, R.C.,   187
Nicholson, J.P.,   231
Noll, E.D.,   56

O'Donnell, J.,   75
Ohtsubo, J.,   173
Olechna, D.,   110
Oliver, B.M.,   162
Oliver, P.,   220
O'Regan, R.,   259
Osterberg, H.,   259
Ostrovsky, D.B.,   262
Owen, T.C.,   247

Padical, T.N.,   87
Page, D.N.,   77
Palmer, E.V.,   190
Pandya, T.P.,   236
Parker, B.L.,   249
Parker Givens, M.,   236
Parks, J.A.,   247
Parma, E.M.,   138
Parrent, G.B., Jr.,   159, 200
Parsons, J.F.,   231, 247, 259
Patorski, K.,   217
Pedrotti, L.S.,   6, 8, 190
Penney, C.M.,   179
Penz, P.A.,   206
Pernick, B.J.,   84, 220
Phalippou, D.,   213
Phillips, N.J.,   229
Phillips, R.A.,   220
Pierce, D.T.,   212
Pipan, C.A.,   220
Pipes, P.B.,   190
Pitlak, R.T.,   159
Pollard, P.,   143
Pontiggia, C.,   140, 183
Porreca, F.,   212
Porter, A.B.,   220
Porter, A.G.,   222
Porter, D.,   229
Porter, W.H.,   41
Powell, R.L.,   259
Protter, M.H.,   63
Punis, G.,   75

Porcupile, J.C.,   44
Pusch, G.,   48

Ragnarsson, S.,   166
Redman, D.,   220
Rees, A.J.,   58
Rigden, J.D.,   162
Rinard, P.M.,   191
Rinkevicius, B.S.,   179
Robbins, P.J.,   87
Roblin, M.L.,   233
Rodemann, A.N.,   166
Roedel, K.H.,   262
Rogers, G.L.,   201, 231, 236, 247
Rome, T.L.,   122
Rosenberg, S.,   179
Rosenfeld, J.,   190
Rottenkolber, H.,   138, 148
Rowe, D.S.,   179
Rubin, D.,   110
Rudd, M.J.,   179
Rudolph, P.,   249
Russo, V.,   229
Rust, D.C.,   24

Sadowski, H.,   151
Sakharov, A.N.,   87
Sandilands, S.,   143
Sardesai, P.L.,   122
Savir, D.,   182
Sawatari, T.,   173
Schadt, M.,   107
Schawlow, A.L.,   115, 201, 202
Schiekel, M.F.,   106
Schober, H.,   185
Schubert, W.K.,   249
Schwar, M.J.R.,   236
Scudieri, F.,   164
Selby, P.H.,   63
Selvarajan, A.,   255
Sen, D.,   151
Sette, D.,   164
Sever, G.A.,   213
Shiffrin, K.S.,   87
Shustin, O.A.,   138, 255
Sigler, R.D.,   132
Silva, D.E.,   216, 217
Singh, M.,   250
Smartt, R.N.,   154
Smith, H.M.,   229
Smith, J.H.,   156
Smith, W.H.,   125
Smith, W.J.,   112
Snyder, D.,   179
Solarek, D.J.,   115

SUBJECT INDEX:

""""""""""""

NOTES :